Praise for

Pain and Prejudice

A scientific treatise, a page-turner, an exposé. It's hard to exaggerate the attractions of this extraordinary book. It makes the personal political and the political personal, drawing the reader along in the careful and scientific exploration of the sexism, biases, and silences of science. *Pain and Prejudice* should be required reading for all scientists.

— Pat Armstrong, Distinguished Research Professor, Department of Sociology, York University, and Fellow of the Royal Society of Canada

How can scientists be objective and empathetic at the same time? Karen Messing's decades of research into workers' health, especially the health of women workers and those of the lower rungs of the working class, are examined and analyzed in a very interesting and readable style. Dr. Messing shows how collaboration with community partners such as unions can improve research but how this type of research is increasingly threatened. She shows how research can and should make change in the workplace to improve workers' health.

— Cathy Walker, past director, National Health and Safety, Canadian Auto Workers

Karen Messing is a riveting storyteller who illuminates areas usually enveloped in the fog of expertise and pedantry. She belongs to a lamentably rare breed; she is a militant intellectual. An accomplished scientist, she tells, in a personal, evocative style, of the way she came to better understand the relationships between employers, science, and labour. Her encounters with, and analyses of, science and scientists hired by capital and government to regulate working conditions lead her to question both the impartiality of science and the accompanying lack of empathy for workers, particularly women. This is a valuable book for anyone interested in social theory, sociology, and, most importantly, the health and safety of workers.

— Harry Glasbeek, author of *Wealth by Stealth*

Messing has long been one of the leading practitioners of "listening to workers' stories" as a way of understanding their health. *Pain and Prejudice* describes how this approach evolved, why it is so effective, and some of the leading findings. It provides a unique window into the world of worker health and safety.

— Wayne Lewchuk, professor, School of Labour Studies and Department of Economics, McMaster University

Pain and Prejudice

What Science Can Learn
about Work from the
People Who Do It

Karen Messing

Between the Lines
Toronto

Pain and Prejudice:
What Science Can Learn about Work from the People Who Do It

First published in 2014 by: Between the Lines
401 Richmond St. W., Studio 277
Toronto, Ontario M5V 3A8
1-800-718-7201
www.btlbooks.com

Library and Archives Canada Cataloguing in Publication

Messing, Karen, author
Pain and prejudice : what science can learn about work from the people who do it / Karen Messing.

Includes bibliographical references and index.
Issued in print and electronic formats.
ISBN 978-1-77113-147-6 (pbk.).
ISBN 978-1-77113-148-3 (epub).
ISBN 978-1-77113-149-0 (pdf).

1. Industrial hygiene. 2. Radiation – Toxicology. I. Title.
RC967.M48 2014 613.6′2 C2014-902457-6
C2014-902458-4

Cover design by Jennifer Tiberio. Cover photo © iStockphoto.com/angelhell
Text design and page preparation by Steve Izma
Printed in Canada

Between the Lines gratefully acknowledges assistance for its publishing activities from the Canada Council for the Arts, the Ontario Arts Council, the Government of Ontario through the Ontario Book Publishers Tax Credit program and through the Ontario Book Initiative, and the Government of Canada through the Canada Book Fund.

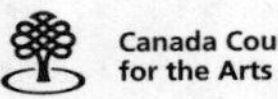

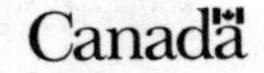

Contents

Preface

MANY RESEARCHERS IN OCCUPATIONAL HEALTH never actually get the chance to talk to people suffering from the work-related health problems they study; their research keeps them in their laboratories, far from the factory floor. But for the past thirty-seven years, I have been lucky enough to be forced into direct contact with the world of work and made to see, hear, smell, and touch the environments that make workers sick.

When I arrived at the Université du Québec à Montréal (UQAM) in 1976, its Department of Biological Sciences had only existed for seven years. The political excitement of the 1960s and the "quiet revolution" had inspired the province of Quebec to create a publicly funded university. By the time I was hired, some professors and administrators had persuaded the university that it should do something for "communities not traditionally served by universities," meaning unions, women's groups, and community groups. After some negotiations, UQAM promised to pay professors to do research on topics suggested by these groups. UQAM created a community outreach service and even hired co-ordinators to link professors with community needs.[1]

One day, the co-ordinators visited our department and asked whether anyone was interested in being a resource. My collaborator and friend Donna Mergler, a professor of physiology, encouraged me to participate; she was already giving educational sessions on the health risks of noise and asbestos. I couldn't see how my doctorate in molecular genetics of lower organisms could be directly useful to the community, but I put my name down. A couple of months later, the co-ordinators called to tell me they had gotten a request for help from radiation-exposed refinery workers who needed a geneticist. My career took an irreversible turn.

Over the following years, Donna and I developed a research program in occupational and environmental health that benefited from community input. In 1990, we founded CINBIOSE,[2] a multidisciplinary research centre that has been able to support other community-friendly researchers with similar interests.

Stimulated in part by the program at UQAM, Marc Renaud, the new head of a provincial government organization that gave grants for health research,[3] decided to offer a program unlike any other source of support for scientists. The grants would be given to university-community partnerships on presentation of a joint research program. In order to ensure that the scientists would listen to the community group, the group or a co-ordinating organization would control the money. The peer review committee rating the proposals would also have community representation.

CINBIOSE got a call from the community outreach office suggesting we apply. For the next fifteen years, until the program was abolished, we got large amounts of money to partner with the women's committees and health and safety committees of Quebec's three largest trade union confederations. Our original partnership included ergonomists, sociologists, and legal scholars, as well as the six union representatives. We called it *l'Invisible qui fait mal* (literally, *The Invisible that Hurts*), referring to the fact that occupational health risks in women's jobs are often less impressive and obvious than they are in men's jobs. This group sponsored dozens of research projects and interventions. We created a book on ergonomics and women's work that European unions translated into six languages and we wrote a United Nations policy paper on gender and occupational health.[4] Our legal specialists helped get new laws passed and old ones respected.

I have been surrounded by a strong, active, nourishing support system, favourable to labour, while doing research in occupational health. This explains why, even while in academia, I have been able to listen to workers' stories and helped to understand them. I have been allowed to observe how lack of respect and understanding from employers, scientists, and the public affects their working conditions and thereby their health. I have been freed from some of the constraints of the scientific establishment and allowed to see how scepticism towards workers' pain has influenced and even shaped the academic field of occupational health research.

The context in which we were able to develop these relationships is fast disappearing as the globe veers to the right. Canada's conservative government has replaced community representatives with industry spokespeople on research granting organizations and "peer" review boards. The Faculty of Science at UQAM expelled CINBIOSE from its sponsored research centres and sent us off to the Faculty of Communication. Our provincial funder closed its doors and its successor decided that our publications were not being seen in the right academic venues. So it will not be as easy for those who come after us to cross the gap between the university and the community of low-paid workers, which I call the "empathy gap" – an inability or unwillingness among scientists and decision-makers to put themselves in the workers' position.

I retired from teaching in 2008, although I continue to do research and supervise graduate students. In this book, I will explain some of what I learned from workers about their jobs, their health, and their lives. I will try to demonstrate how the gap in experience and interests between low-paid workers and the classes above them affects their health and the scientific discourse about it, with costs for scientific quality, for the public, and even for employers.

I will describe how I learned about the reality of work at the bottom of the social hierarchy and was confronted by how little employers and the public know about that reality. I came to understand how the science I had learned in school and from my colleagues was often inadequate to interpret the effects of low-paid jobs on workers' health. And I will try to show that, because of its ignorance of many real working conditions, the science of occupational health does not always do as good a job as it could in finding out about all workers' health.

Toward the end of the book, I try to examine my academic field a bit more critically. Do occupational health scientists just lack empathy with people of lower status? Or are there more structural problems at work? It seems to me that there is a complicated interplay between scientists' training, their "personal" attitudes towards workers, and their economic and social interests. I have no expertise in political science or economics to help me tease out these relationships, but I do know that scientists could be more helpful than we are in improving workers' health. My last chapter tells some success stories that can suggest paths for change.

But before discussing solutions I have to explain how I experienced the gap between workers and scientists and how my colleagues, students, and the workers I came to know pulled me across that gap.

This book owes a lot to my students and collaborators, from whose work I have freely borrowed. They will be cited in the text, but I want to mention specifically: Donna Mergler, Ana María Seifert, and Nicole Vézina, who showed me how to do occupational health research with respect for workers; Katherine Lippel, who made me aware of how complicated occupational health policy is; Stephanie Premji and Jill Hanley, who led me to the story of the many ways that immigrants' health is affected by their specific workplace conditions; France Tissot, Stephanie Premji, and Susan Stock, who have worked with me on the hidden inequalities in epidemiological analysis; Carole Gingras, Lucie Dagenais, Jocelyne Everell, Céline Charbonneau, Nicole Lepage, Micheline Boucher, Pierre Lefebvre, Marie-France Benoit, Ann Potvin, Ghislaine Fleury, Gisèle Bourret, Michelle Desfonds, Sylvie De Grosbois, Martine Blanc, and Sylvie Lépine, among others, who helped me again and again to understand the needs of the workers in their unions; and Florence Chappert of the Agence nationale pour l'amélioration des conditions de travail (France) who gives me hope. I also sincerely thank (in no particular order) some of the courageous worker-friendly scientists who have encouraged and inspired me: Barbara Silverstein, Laura Punnett, Hester Lipscomb, Céline Chatigny, Cynthia Cockburn, Patrizia Romito, Annie Thébaud-Mony, Laurent Vogel, Maria De Koninck, Jeanne Stellman, Jim Brophy and Margy Keith, Romaine Malenfant and Robert Plante, Ghislaine Doniol-Shaw, Danièle Kergoat, Catherine Cailloux-Teiger, Ruth Hubbard, and Åsa Kilbom. And the many generous workers who let us observe them and explained what we couldn't see.

More specifically I owe thanks to Melissa Wakeling, Glanmore National Historic Site, Belleville, for the reference used as an epigraph for chapter 7, to Donna Vargas for information on the Montreal Day Nursery (chapter 2), and to Chantal Lavigne of Radio-Canada for generously sharing her research results on teachers (chapter 8).

I was lucky enough to be encouraged by the Scribblers: the late Martin Kevan (we miss you), Barbara Scales, Ana María Seifert, and Pierre Sormany. Pat Armstrong and her graduate students at York University were very helpful, especially Suzanne Day. Katherine Lippel, Harry Glasbeek, Cathy Walker, and Daood Aidroos gave very helpful comments on parts of the book. Amanda Crocker, my editor at Between the Lines, and copy editor Cameron Duder both made important, thoughtful suggestions for improving the manuscript. I didn't take everyone's suggestions, but I appreciated all of them. Thanks to Gloria Steinberg for keeping me going, and to my wonderful family, in-laws, and stepfamily for their love and support. And of course to Pierre for input and support with the book, the title, the research, and my life.

Chapter 1

Factory Workers

When I was little, my father took me for a morning to the factory where he was an executive. To my delight, he let me sit at the line and watch the women wiring radios. The red, blue, and yellow wires had to be soldered in the right places in each radio. The women even let me play with the coloured wires while my father was busy. This occupied me for a while, but then I got down off my chair and went to see my father in his office. I had something on my mind. I asked him, "Don't they get bored doing the same thing all day?" He replied, "No, they don't. They're not smart like you, Karen."

I was floored. My father was telling me that these grownup women were not as smart as me, a five-year-old who had a pretty good idea of my low rank in society. What he was saying didn't seem too plausible, but he seemed to be sure of what he said. I puzzled over this for a while and never forgot it.

Many years later, circumstances conspired to suggest to me that my father might have been mistaken about the intelligence of workers. When I was seventeen, I was suspended from my university for a piece of minor mischief and it would be three months before I would be allowed to go back. I applied for jobs in a bookstore and several restaurants and was finally hired as a waitress at a cafeteria known for its quick lunches. I was supposed to supply each customer with a tray, napkin and silver, take the order and yell it to the kitchen staff with the right code name, in detail, for each preparation (special hold the green, burger New York . . .). For each of the ten or so main dishes, I had to supply the right side dishes or condiments. If the order appeared with all its fixings in due course in the window from the kitchen, I had to give it to the right customer. If it didn't, I had to negotiate with the kitchen staff, trying to balance the customer's grumbling against the way Henry the cook, a scary guy, would get annoyed with me for nagging him.

I was a pretty terrible waitress. The women who had been doing counter service for several years were able to handle orders from four customers at a time. Slap! Slap! Slap! Slap! went the trays on the counter and the silver and the food on the trays. But I never could manage juggling more than two customers. And – most humiliating for an honours Ivy League student – the biggest obstacle was not physical, it was mental. For the life of me, I couldn't manage the cognitive challenge of getting the orders and their details right and following their progress for more than two customers at a time. Beverly, a girl my age who had been hired just before me, was a great comfort. She explained little tricks she had picked up, like forgetting about the parsley on the egg if there was a long line of customers. And it was getting to know Beverly that put the final touches on my growing suspicion that people could be working class and bright at the same time. She had all the stigmata of a life in poverty – missing teeth, uneducated speech patterns, sole support of a new baby – and she was at least as quick mentally as I was. We had a great time together making fun of the managers and the kitchen staff until I went back to university and my real life.

It was also in this job that I began to understand about power relations between employers and workers. Beverly and I were paid what was then minimum wage, $1 an hour. To me, even in 1960, this seemed a tiny amount of money, and I couldn't understand how Beverly and her baby could live on it. Especially since the employer made us pay him for cleaning our uniforms. It seemed to me that we shouldn't have to pay for this since the uniforms didn't belong to us. But the manager quickly made me understand that if I wanted the job, I would pay for the cleaning and I would shut up. And a few of the customers were quite as efficient in getting across the idea that if I wanted to keep the job, I wouldn't object to any of their patronizing or flirtatious remarks. If I smiled enough, sometimes they would even leave me a quarter on their tray.

For the next few years, I behaved myself and finished university and graduate school. My only close encounters with low-paid workers took place as a customer. It wasn't until I got a job as a biology professor at the Université du Québec à Montréal (UQAM) that I had another, very different kind of contact.

In 1978, a problem arose at a phosphate refinery near Montreal. The men who worked there heard that the ore they processed was contaminated with radioactive dust. The refinery's waste rocks had been sold to the province as road paving, and a technician had noticed that the roads emitted radiation and worried that commuters might be exposed. Reading the newspapers, workers in the plant learned for the first time that the material they handled was radioactive and dangerous to human health. They called the union, who called the university outreach office, who called me. I was the only potential resource who came close to knowing anything about radiation and genetic damage. The union health and safety counsellor and I drove out to the south shore of the Saint Lawrence River one cold day and met with the union executive in their little union office. There were six men in their thirties and forties who had spent years in the plant. They told us that not only was their workplace full of radioactive dust, but the workers had been taking home the factory waste to use as phosphate fertilizer in their gardens. I knew little about the effects of radiation on humans, but I gave the six men the Genetics 101 version of how radiation works: its energy can damage chromosomes and thereby change genes, and altered genes don't work as well, possibly affecting health. I glibly mentioned that the damage could be passed on to the next generation and beyond.

"So my daughter's problem could come from my job?" asked the union president "Jean-Jacques." With a bump, I woke up to the fact that I wasn't in my classroom and should have been gentler. Too late – I had created a shock. Of the six men around the table, five were married, four had children, and those four each had a child with a significant health problem, from cleft palate to clubfoot. The fifth married man had a pregnant wife, and he and, suddenly, I were both worried about their future child. And, yes, several months later the child was born with a serious birth defect: she suffered from tracheal-aesophogeal fistula, a condition where there is a hole between the air passage and the digestive tube running from the mouth to the stomach.

I had no idea how to approach human genetics professionally, but it was clear to me that someone had to do something to find out whether there was a problem at the plant. Thus began a frustrating and mystifying period where I tried to contact qualified people – university professors and medical researchers – and interest them in helping those hundred men exposed to radiation to find out what was happening to them and their families. Mystifying because, for some reason, none of the logical people to contact showed any interest in getting involved in a situation that was, to me, humanly compelling as well as scientifically fascinating. I first called a genetics researcher at a children's hospital in Montreal and, in my innocence, started out on the wrong foot: "I'm Karen Messing and I'm a biology professor at UQAM and we have an agreement with a union to give them information on health and safety risks and we need an expert on human genetics." "No, I'm not interested in working for a union" was the reply. "No, I didn't mean the union would hire you, it's just that these people are exposed to radiation and have malformed children and I don't have the expertise to judge whether the radiation is causing the problem," I explained. "No, I'm not interested in working for a union," he repeated.

One of the union executives whose wife had had a malformed child was referred to a local hospital genetics counsellor I will call "Dr. Tremblay."[1] Dr. Tremblay told him, "These things just happen, we'll never understand them. But they could not be associated with your work." I heard about this and tried to reach Dr. Tremblay to find out why he thought they could not be associated with the executive's work. I left messages for him and then for others in his service, but no one returned my calls. And so it went, even though I edited any mention of the union out of my subsequent phone calls. None of the people whose job it was, none of the researchers whose expertise it was, would meet the workers as a group or study their situation. Just imagining a potential conflict with an employer was enough to put off my colleagues, who, to do them justice, hadn't met the distressed fathers. And truly, talking with them was maybe best avoided. I still haven't forgotten the face of the man who said, "I worked all my life in this crummy plant to keep my family safe and healthy and now you're telling me I maybe gave my son his heart problems." Or the fiancée of another worker who explained to me that she had broken off her engagement because she wanted to have children and was afraid of radiation damage.

At that time, I had just joined our biology department and my research program in genetics was aimed at developing and strengthening a fungus that would kill mosquitoes. I had gotten a grant with two of my colleagues, entomologists who knew how mosquitoes should be killed, and we were doing well. I had hired some students who were busy growing fungus on plastic dishes and floating their spores on the surface of water where the mosquito larvae lived. My department was happy it had hired me because I had shown I could get grants from federal and provincial sources.

What was I going to do about the refinery workers? I talked it over with Micheline Cyr, Ana María Seifert, and Claire Marien, three very bright undergraduate biology students who were looking for a term project. They offered to look at the radiation exposures with me. Week after week we read about radiation and discussed how to deal with the scientific and human issues at the factory. We felt terrible because we had no way of knowing whether the children's problems were in fact due to their fathers' work, and no one would help us find out. We didn't know whether to reassure the fiancée or to commiserate with her.

We were especially concerned because we were learning more about the working conditions at the plant. We met a veteran worker who told us how dusty it was. He explained to us that when the workers needed to have dental work done, their dentist made them stop working for weeks before; otherwise their jaws were too weakened by exposure to the phosphorus in the dust, and the dentist was afraid to damage them. (A year later, when the employer allowed us to tour the plant for the first time, every surface was dusty. We felt dusty ourselves, inside and out, after just a few minutes.)

The union arranged a Sunday morning breakfast with the men and their families so they could fill out a questionnaire on their reproductive histories, to give us some idea whether there were too many miscarriages, stillbirths, or malformed children. Micheline, Ana María, Claire, and I went to a small church basement where the wives served us bacon and eggs and pancakes. I gave a presentation about the effects of radiation on health. This time I was a bit more tactful but the audience was understandably tense. Still, the wives were glad to have a place to ask their many questions. Most of the workers were too young to have had children and so only thirty wives filled out the questionnaire. Their answers confirmed that there might be a problem but we couldn't really be sure because there was such a small sample to work with.

The scientific literature was not much help to us because it only talked about rays. It was full of calculations about the doses people would get from an external source of gamma or X-rays, but I couldn't find anything about what happened if you swallowed radioactive dust. How long would the dust stay inside you? Would the radioactive elements concentrate in certain organs? Given that the radiation was not outside the body, all the calculations in the scientific literature about the distance of the human body from the radioactive source were no use to me. So we needed some way to find out whether the radiation was causing damage.

I had been to school with Professor Abby Lippman of McGill University, who had done her Ph.D. in genetic counselling. She was able to introduce me to a clinician who knew something about how to examine human chromosomes. Dr. Naomi Fitch kindly offered to teach me how to examine the chromosomes myself, so I went to her lab for re-training. (This was at a time when scientific institutions were less rigid; today, no department or funder would give a newly hired professor enough slack to dabble in extraneous research fields.) The union was able to access a little money via the UQAM-union agreement, enabling me to hire Micheline, Ana María, and Claire to help out with the chromosome study. We collected blood from a small sample of workers and prepared slides with special staining to be able to see the chromosomes. The students and I examined the slides (which had been mixed in with slides from other people so they could be classified "blind") and we all thought we saw many more signs of damage in the samples from workers than those from people who didn't work at the refinery.

When we sent our report to the union and the employer, some other geneticists finally got interested. In fact, I got a call from the president of the Canadian Genetics Society asking for my slides – he had been consulted by the employer and wanted to do a counter-expertise (a paid contrary opinion). And, after more than a year of silence, Dr. Tremblay the genetics counsellor finally called me back. He told me the employer had also consulted him and had told him to threaten me with a lawsuit. He warned me to drop the whole idea of doing a study.

Further education about the workings of North American science followed. The students and I were still worried about our lack of experience in examining human cells and wanted to have independent confirmation of our results from someone not associated with the union or the employer. A friend of a friend was a recognized expert on occupational health in the U.S., a professor at a major school of public health. When we contacted him, he offered to re-do our study. We were delighted, since we needed to know whether it was true that the refinery workers' chromosomes had been damaged. "Professor Ivy" came, took blood samples, and disappeared from view. A few months later, all the (French-speaking) workers received a letter in English on the letterhead of his prestigious university saying their chromosome tests were negative and they had no problems. We did not get a copy, but the union called us and said our results had been wrong. When I phoned him for confirmation, Professor Ivy told me that sending the letters was a secretary's error and that he had not yet looked at the samples. He had been busy with other things; he was negotiating tenure at his job. He offered to send a letter of correction to the workers but never did that either, resulting in further confusion. During our phone call, he did ask me for permission to use a photograph of one of the workers' blood samples in a textbook he was editing, because he had never seen a worker's chromosome damaged in that particular way.[2]

We never received any information about his analysis of those samples, so we never knew whether our results had been accurate, although Professor Ivy's interest in the highly damaged chromosome seemed to indicate that there might be some problem. More importantly, to this day, over thirty years later, the workers have never been given any confirmation about their health status or the future of their children and grandchildren. Because of the episode with Professor Ivy, they were never sure they could trust us either. In light of all the conflicting information, the best the union could do was to bargain with management, offering to stop dealing with us if management would clean up the radioactive dust in the plant and install a new ventilation system. Management was happy to accept, and we were ousted. As far as we know, no other scientist has ever studied these workers' chromosomes. The refinery closed in May 1992, leaving behind 1.2 million metric tons of radioactive waste to be disposed of by the public health authorities.[3] Professor Ivy went on to head a U.S. government occupational health research organization and still has an excellent reputation in scientific circles.

My brush with radiation science was not over. During our study of the refinery workers I had read and learned a lot about radiation and its effects on human beings. I also got connected with a loose network of researchers in the new field of occupational health. In the 1970s and 1980s, people were just starting to become aware of industrial contamination. Dr. Jeanne Stellman of the Oil, Chemical and Atomic Workers Union in the United States had recently published *Work is Dangerous for Your Health*, which alerted workers to their chemical and physical exposures.

Two years later, because of my growing familiarity with radiation exposure, I accepted when a hospital union asked me to be an expert witness for an X-ray technician who was asking her employer for paid leave for the duration of her pregnancy. "Suzanne" had already had a malformed child, and several of her colleagues had recently had miscarriages. She didn't want this fetus to be exposed to radiation and had left her job on her own initiative. She was trying to get back pay through a contract provision allowing for leave for pregnant women exposed to workplace risks.[4] The employer maintained that she was not exposed to any risk because her dosimeter readings showed very low exposures.[5] But the union estimated that the risk of accidental exposure was unacceptably high because workers said that the prescribed precautions were frequently ignored. Workers had seen medical students turning on X-ray machines before the technician had left the room, and they had counted the number of times the lead-shielded doors were inadvertently left open. They also didn't trust the dosimeters because some technicians had purposefully left their dosimeters in front of the X-ray machine for long periods without ever hearing that their readings were abnormal.[6]

I thought the union had a chance of winning the grievance, but I did not realize what we were up against. At that time, in 1980, I was an untenured professor, thirty-seven years old. My only published research was on fungus. Although I had done a lot of reading and research on radiation, it would be five years before I would publish my first peer-reviewed scientific article about human genetics – the most important qualification for expertise. As I prepared for the hearing, I found out to my consternation that the employer's expert was Professor Edward P. Radford. He had chaired an international commission on the effects of radiation on humans and had published hundreds of peer-reviewed articles and communications on the subject. Debate was currently raging in scientific circles about the effects of low-level exposures to ionizing radiation and Radford was among those who perceived those effects to be negligible. I felt scared and inadequate. I tried in vain to get other, better qualified scientists to replace me, but no one would accept going up against Radford, not even when the union offered to pay them at an appropriate rate for international experts.

The hearing, which I remember as taking place in a little dark room in Quebec City, was tough. A small army of lawyers had been paid to show I was wrong. Although there was a superficial resemblance to my doctoral defence, where three professors attacked my results, the stakes were much higher at this hearing. I felt naively as if my competence alone would determine the fate of the children of the radiology technicians, whom I had come to know and care about. I thought that if I did well, their future fetuses would not be exposed to radiation. Our side had no lawyer, just a self-educated union counsellor, and I was unprepared for the kinds of questions I got from the opposing lawyers. Scientists have no training or experience in how to deal with legal issues and I was no exception. I had discovered a few publications that supported our argument that technicians could be exposed to dangerous amounts of radiation, and I summarized them. Even though Radford had gone back to the U.S., I knew that he would get a translation of my testimony and could tear it apart if I made a mistake. Still, I felt that Suzanne had a good case, since there was a proven risk of an accidental exposure and the union had raised doubts about the dosimeters. Also, even if the scientists had been able to agree that there was no danger from the radiation, wouldn't Suzanne be exposed to an inordinate amount of stress during her pregnancy, given her belief that her previous child had been damaged by her work?

We lost, of course. It wasn't even a contest. The judge decided that the union's testimony about actual workplace practices was not relevant and concluded that the technician's job was "no more dangerous than being exposed to sunlight."[7] I felt humiliated and guilty for not having been up to the task. Luckily, Suzanne's baby, born long before the verdict was pronounced, seemed perfectly healthy. Suzanne, who had quit working at her own expense, was able to go back after childbirth with no ill effects, other than loss of income. But there would be no leave for pregnant radiology technicians.

I now think that my experience at the hearing explains a lot about why my more experienced, older colleagues had shunned the refinery workers. Getting a Ph.D. requires scientists to learn thoroughly how to do one kind of scientific research and respond to one kind of critique. We become expert in responding to other scientists' questions, and we are able to anticipate them and do our work carefully to avoid criticism. Our whole sense of self depends on being right, leaving no stone unturned, no argument unanswered. We learn the rules of scientific discourse and some packaged statements to produce when we are not sure. Uncertainty on our part is OK, even praiseworthy, because we are admitting we don't know all the answers and are still seeking the truth. In fact, for scientific articles, the standard says we are only allowed to make a statement if we have less than a

5 per cent chance of being wrong. The punishment for not observing the rules is harsh: our thesis will not pass, our articles will be refused, our grant requests will not be funded, we will get nasty, critical reviews from our colleagues, people will make fun of us in conferences, we will not get jobs, students will not want to work with us. So our whole, very long period of training teaches us never to say anything positively. We are taught to feel contempt, even disgust for scientists who make unqualified positive statements. They are not being scientific.

When science intersects with real life, the rules change. In courts and government hearings, scientists are supposed to give opinions. I remember U.S. Senator Ed Muskie who listened to a series of scientists testifying on environmental hazards and said mockingly, "On the one hand this is the case, and on the other hand that is the case. . . . We need more one-handed experts who can be explicit on what should be done." The judge at Suzanne's hearing needed to make a decision, so any shillyshallying was unwelcome. In law, an expert has to go with the most likely interpretation – to live with up to a 49 per cent probability of being wrong. And Suzanne needed me to explain clearly why I thought her fetus was in danger. Protecting my scientific reputation by hedging was not an option. So, at the hearing, I was repeatedly put in the unfamiliar position of having to make bald, un-nuanced statements that reflected my best guess. And I was faced by a heavy opposition that was having no problem telling me I was wrong. I felt off-balance.

So other scientists are not necessarily indifferent to workers' anxiety and suffering – just uncomfortable with making statements that go counter to their training and understandably unwilling to undergo stress and humiliation in a lost cause. Even reluctant to get too close to the actual people whose lives are affected by their research, for fear of losing their "objectivity."[8]

If scientists testify in any arena where the money is all on the opposing side, they can be sure that they will be subject to a barrage of questions and criticisms. Even though they may believe in the case they defend, they will be well aware that not every i has been dotted nor every t crossed. They will have an unacceptable choice: make strong, positive statements that overstate their perception of the evidence, or qualify everything that they say, making their testimony useless to the sick or endangered worker. I had accepted the union's request to testify only because our university's unique agreement with the trade unions and community groups had given me up-close access to workers and union counsellors. I remembered the anxiety of the refinery workers. As a mother, I empathized with Suzanne's terror about having another malformed child. And I deeply respected the devotion of the union health and safety counsellors. But most academics don't have any chance to meet these people, and no way of understanding their experience. In the thirty-two years since the hearing, I have become aware of a gulf in experience that separates low-status workers from academics and others in the social classes above them, and workers don't have access to media where their experience can be publicized.

The cost of the gulf in experience or "empathy gap" is enormous – to working people, to science, and to society. When judges who decide on workers' compensation can't imagine conditions on the assembly line, they refuse workers' claims for compensation for work-related illness.[9] When scientists don't understand why workers' feet hurt, they design studies with flaws that make them appear to show that workers' feet don't hurt. When employers don't realize that hospital cleaners really know a lot about cleaning and care about making their workplace shine, they don't consult them before buying equipment and furniture that make them inefficient and even ill.

Of course, the gulf in experience is not the only reason academics shy away from workers. I suspect that Professor Ivy might not have been so cavalier with our slides if he had had a contract to analyze them with an international corporation. Empathy, to some extent, can be bought, and some international corporations have become expert in getting scientists to empathize with their point of view. I think it must be easier to force yourself to express a scientific opinion without nuance if the army of high-ranking lawyers is on your side rather than being lined up against you.[10]

Still, the radiation-exposed workers had a lot going for them: they were respected members of a small supportive community, they had a relatively strong union with active support from a technically savvy union service staff, they were exposed to radiation, a scary-sounding risk, and the public had been alerted to the risk of radiation so the employer was under scrutiny. Although they may have been unable to protect their health completely, the workers were at least able to get a new ventilation system. In the next six chapters I will describe people who have few advantages to help them gain support for change.

Chapter 2

The Invisible World of Cleaning

They'll never see God in their whole lives, for making that black.

— *Cleaner in a Quebec hospital, dusting the new shiny black furniture that showed every speck of dirt*

I KNEW NOTHING OF ERGONOMICS when I became a biology professor at UQAM. It was only through many encounters with hospital workers, especially with cleaners, that I travelled from genetics to ergonomics. (I will explain more about ergonomics below, but for now I will just define it as the analysis of paid work activity in order to improve it.)

Discouraged after our adventures with radiation-exposed workers, my students and I went back to my research into the genetics of a fungus that kills mosquitoes. I had worked with this fungus during my postdoctoral fellowship, and it seemed to me that developing non-chemical approaches to killing mosquitoes would be good for the environment and satisfy my need to feel useful. I had two excellent students doing experiments to find out how the fungus blocked the mosquitoes' breathing tube and how we could help its genes do better at killing the mosquitoes. However, the union-university agreement had not finished with me.

After our failure to gain leave for the pregnant worker, my contacts in the radiology technicians' union wanted to know for sure whether the radiation they were exposed to would make them miscarry or have babies with health problems. Even though their dosimeters showed very low amounts of radiation, I thought it would be worth looking to see whether technicians' babies were more likely to be unhealthy than those of other health workers, say nurses. Professor Abby Lippman, an epidemiologist friend, helped us produce a questionnaire about radiation exposure and pregnancy outcomes. Ana María Seifert, the Bolivian refugee who had helped examine the refinery workers' cells, joined the technicians' project as a research assistant. I was lucky to have Ana María in my laboratory since all the plastic pipettes, test tubes, and flasks we were throwing away after a single use scandalized her Bolivian soul. In 1980, way before anyone else, she made us become early adopters of environmentally sensitive laboratory practices like using glass instead of plastic. Her warm, friendly nature made her a real help, too, when we were meeting with health care workers to develop a questionnaire on reproductive problems.

Abby insisted we pre-test the questionnaire by getting the hospital workers to fill it out in our presence, so we invited a group of nurses and radiology technicians to the university and spent a few hours with them.[1] The more Ana María and I listened, the more we realized that hospital workers were exposed to a lot of conditions that might hurt fetuses, not just to radiation. We would also have to ask about lifting weights, stress and tension, uncomfortable postures, and exposure to germs and hazardous chemicals. The questionnaire became quite long and the union had some trouble finding enough people who could take the time to fill it out.

Because not all workers become pregnant (fewer than 2 per cent of working women per year) and problem pregnancies are relatively rare events (3 to 20 per cent of pregnancies depending on what you count as a problem), we needed a lot of them to answer the questionnaire. We needed to be able to tell whether the number of reproductive problems found among the technicians exposed to radiation was really greater than those among technicians before they were exposed to radiation or among nurses much less exposed to radiation, or whether any differences were due to chance. If we had only a few respondents, we would cover only a few pregnancies, not enough for statistical significance. In the end, we didn't succeed in getting enough. The results we had didn't show any difference between nurses and radiology technicians, but this could have been because the number of pregnancies we had information on was too small.[2]

But it is also possible that we couldn't find any difference because nurses and technicians were both exposed to other conditions that are dangerous for fetuses and swamped the radiation effects. We were quite surprised by the number of health risks we identified in the jobs of these hospital workers. At that time (the mid-1980s) lifting equipment was scarce and was hard to access. Moving a 300-lb weight is considered risky for strong young men, but many heavy patients have to be moved by mostly female hospital personnel. Workers rarely had the time to wait for another person to help them, so both nurses and technicians lifted and moved around a lot of patients alone, not the best activity for pregnant women. In fact, heavy lifting is significantly associated with pregnancy complications, as are other conditions to which hospital workers are exposed, like prolonged standing.[3] We asked ourselves whether the major danger to pregnant women in hospitals could come from the physical demands of work, not from the radiation, and the workers themselves asked us whether the intense work speed and "stress" could affect their pregnancies.

From the questionnaire responses, we did get information on the fast work speed, frequent lifting, awkward postures, and other physical elements of the workers' jobs, as well as some risks of accidental exposure to radiation. We had a series of meetings with radiology technicians all over the province to tell them about these results. We met with them in hospitals and union headquarters, gave them a report on what we had found, and listened as they vented about the changes going on in their jobs that made the physical challenges even more intense. The government had recently decided that patients shouldn't be in hospital unless they were very sick, that most patients would be better off being cared for at home. Hospital stays were to be very short and confined to the period when the patient was acutely ill. This apparently benign policy had unexpected effects on hospital workers. For example, when hospital stays are shorter, the number of hospital admissions per day must rise. But the number of admissions clerks stayed the same, so their workload rose considerably.[4]

For the radiology technicians, the new policy meant that the patients they examined were, on average, sicker. Since the patients went home the minute they were able to move around, a much larger proportion of those in hospital were in wheelchairs or needed help with mobility. The technicians were doing a lot more lifting, but there were no more technicians and they had little or no new lifting equipment. They complained that the radiology department budget went instead to fill the examining rooms with the latest innovations in imaging equipment, making it even harder to move patients in the cramped spaces. I emerged from that study convinced that someone should pay attention to the intensified demands on all hospital workers.

At about the same time, the unions were also asking me to help them out with an educational project. Inspired by a growing international movement, Quebec had recently rewritten its occupational health and safety law, and for the first time there were special provisions for pregnant and nursing workers, called "precautionary leave." Those exposed to a danger for themselves or their fetus or nursing infant could ask for reassignment to a safer job or, failing reassignment, could be sent home at virtually full pay.[5] The problem was that few people in the unions knew what conditions were dangerous for pregnant women. They needed someone to interpret the scientific literature and tell them when to ask for leave.

From 1980 to 1985, I participated in over fifty educational sessions with union members from all kinds of workplaces, from factories to teaching. Donna Mergler, who had pioneered these university-union health and safety sessions, told me she always started a session by going around the room to ask the workers to describe their jobs and health problems. Then she would be able to use the real working conditions to illustrate her explanations of how to detect and prevent dangerous conditions. I adopted this practice for the sessions on pregnancy, so I ended up hearing lots of stories about women's working conditions. I was astonished by what I heard. Like most people, I had assumed that women's jobs were "light work,"[6] but I heard about pregnant factory workers cramped over sewing machines and struggling to make their production quotas, pregnant medical archivists bent over from the weight of hospital files, and pregnant laundry workers fainting from the heat.

As a biologist, I was able to appreciate the costs of the workload for the workers' bodies, and I had studied up enough to be able to alert the workers and the unions to some hazards. But I wanted to be able to do more, to suggest remedies, and my studies and reading in genetics told me nothing about that. In fact, the scientific literature then available had very little information about risks for pregnant women and even less about how to adapt workplaces to workers' bodies.

In 1983–84, I had my first sabbatical year, given to improving my skills in genetics. I wanted to get better at detecting the effects of radiation in living workers, so as to be able to see whether prevention efforts were doing any good. I got some money to do the research and spent the year in a laboratory attached to a big Montreal hospital. I learned about how to grow cells in plastic dishes and how to think about mutations in human beings. I was not very good at culturing human cells, which requires a lot of precision and concentration. Ana María and Micheline got much better cell growth than I did. But I was good at meeting with doctors and patients. I was able to get permission to see whether people being treated with radiation, such as those undergoing radiotherapy, had more mutations in their circulating blood cells than other patients (they did).[7]

I spent every day in the laboratory, fascinated by the results we were getting. But at lunchtime I went to the hospital cafeteria where I ran into union people from time to time. The president of the hospital's union, Claudette Carbonneau, had heard of my sessions with pregnant workers and, when she learned I was at her hospital, she called me in for help with quite a different problem.

There was trouble brewing among the hospital cleaners. Their workload had recently increased due to the rise in patient turnover; every time a patient left a hospital bed, the bed and the whole room had to be disinfected, which took fifteen to thirty minutes per room. Sadly, when conditions get harder, the workers' resulting anger is sometimes misdirected. This time the men and women cleaners had turned against each other.

Cleaning at that time was organized into "light" and "heavy" cleaning. "Light" cleaning (originally called "cleaning – women"[8]) was done by women or, occasionally, by men invalided out of other jobs. The job description for "light" cleaners included dusting, washing furniture, cleaning bathrooms, and emptying wastebaskets. "Heavy" cleaning consisted of washing, vacuuming, and polishing floors, and was almost exclusively done by men.[9] One "light" and one "heavy" worker were assigned to each territory. As a carryover from the time that the two jobs had been officially separated according to gender, "light" cleaning paid less.

Claudette and her union were preparing changes to the collective agreement, and one proposal was to abolish this salary difference. The men objected vociferously, on the grounds that their work was harder – too hard for women – so they should be paid more. The women contended that their jobs were harder than the men's. Claudette asked me to hold a discussion with the cleaners on biological differences between men and women in relation to the task demands in cleaning. I accepted, since I taught sex differences in my genetics course and had published a paper on biological sex differences and job qualification.[10] I read all the scientific literature I could find, but I was not at all prepared for the experience to come.

Hospital management had objected to the session on the grounds that it would interfere with the cleaners' work, so we had to hold it during lunchtime, in an inadequate space. My memory of the experience has people squished against each other and standing on chairs to see and talk. I found that it was one thing to explain the biology underlying size and strength differences to students whose interest, if any, was theoretical, and another to try to counter stereotyped notions held by people whose self-respect was at stake. I tried to explain that size and strength differences between the sexes were not absolute, that there were weak men and strong women. I told them that gender differences in job performance depended on the task, the equipment, and the techniques used.[11] They listened politely, but both men and women wanted more than anything to explain how hard their jobs were. Some complained that management put injured male workers into "light work" in cleaning, but it was too heavy for them. The discussion became animated, to say the least. The women had counted how many wastebaskets full of trash they lifted in a day – an impressive number that I remember as 150. The women said that the men got to push machines around to clean corridors while they themselves had to get down on their knees and scrub bathrooms manually. The men countered by describing informal practices that assigned "the mop guy" to almost any job that required physical strength, from holding down a psychotic patient or moving heavy furniture to lifting hugely obese patients alone. One handsome young man with penetrating blue eyes cried: "We're sick and tired of being treated like oxen!"[12] Each group, women and men, had a list of heavy demands made by the employer that justified a higher salary.

The discussion didn't resolve anything on the spot but I did meet Ginette, the young cleaner who ran the union women's committee, and we got friendly. Over coffee from time to time during my sabbatical in her hospital, I learned that the union had agreed to ask for a raise for the women and that they had eventually succeeded in getting it. Some months after I got back to the university, Ginette called me in distress and invited me to lunch again. She wanted to tell me that the hospital had decided to save money by increasing the surface area each worker had to clean. She and her friends were unable to clean 50 per cent more territory in the same time, so they had to cut corners. They had to leave out some areas or some cleaning operations. They were upset because they could no longer feel proud of their work. Some were taking the risk of using chemicals straight from the container rather than the recommended dilution – a dangerous practice – in the hope that they could spend less time scrubbing. Others were trying unsuccessfully to keep to the old standards. But now the employer was letting it be understood that they would be happy to outsource the cleaning to an agency if their own employees couldn't work at the faster pace. Ginette and her colleagues were caught between these threats and their fear that the hospital would be dirty. In a corner of the hospital cafeteria, she burst into tears, saying, "Can't you university professors do anything?" I felt bad, but couldn't justify her faith in university professors. I had no relevant professional knowledge of her workload or how appropriate it was in relation to human capacity. At that time I knew nothing about ergonomics.

On the other hand, I was getting more and more fed up with the world of molecular genetics. Initially, my collaborators from the hospital laboratory and I had been quite excited about publishing the first paper showing genetic mutations in the blood cells of human beings exposed to therapeutic radiation and later, the first paper on mutations induced by radiation in workers.[13] We won some grants from government sources to develop our understanding further, and I got to go to scientific meetings in Sweden and Finland to present our results. But I felt a bit out of place in the Environmental Mutagenesis Society meetings. Most of the other scientists worked entirely with cells in test tubes in the laboratory and were uncomfortable with the erratic behaviour of cells that had stayed in the human body. They often asked me why we were working on cells from human beings rather than cultured Chinese hamster cells, which were much easier and less expensive to study. Every time I met other scientists I had to negotiate across the empathy gap to try to make them imagine the importance of our research to workers. At the same time, although my students were making great progress in cultivating the human cells in test tubes, we were going more slowly than the U.S. laboratories, which had much better resources. I was spending an awful lot of time just asking for money to pay for the expensive material we were using to grow cells.

My colleague Donna Mergler and I got together one evening in 1986 and counted up our grant applications. She had made a lot of applications to study chemical effects on workers' brains, and I had applied to study radiation effects on genes. We found that the further the study was removed from workers, the more likely it was that our grant applications would be accepted. For example, proposals for *in vitro* studies of cells, particularly animal cells, were much more successful than studies of workers.[14] Probably the reason was that the more artificial the environment, the better the conditions could be controlled. It is easier to do the perfect study on Chinese hamster cells, growing demurely on artificial culture media in plastic boxes, than on workers whose diet, behaviour, state of health, and exposures vary from hour to hour. But we feared that results from hamster cells didn't tell us as much about workers as their own cells would.

Going to scientific meetings was scary, since I had to defend our findings against critiques from people with much prettier results. Also, like many medical science meetings, the Environmental Mutagenesis Society meetings were sponsored by commercial interests, and the drug and tobacco companies were very much in evidence. When I presented some results showing that we detected fewer mutations among smokers under some conditions,[15] I was offered money to pursue this and felt ill at ease.

The idea that we should protect workers from pollutants at the worksite, on the other hand, didn't seem to inspire any interest. In fact, one of the American groups was using the same methods as we were, but with a totally different aim that I didn't like. They were sampling blood cells from workers exposed to toxic agents, just as we were, but, instead of using their results to lower toxic levels, they were proposing to use the tests to try to identify workers who should not be hired. Their idea, which earned them a mention in the prestigious journal *Science*,[16] was this: (1) Identify a factory whose workers are exposed to known toxins that induce gene mutations (methylchloroform). (2) Take blood samples from the workers and expose the cells to the same toxin in the laboratory. (3) Count the mutations in each worker's cells and determine which workers have the most and which the least. (4) Wait several years, go back to the plant and see whether the workers who have gotten cancer in the interim are in fact those whose cells made the most mutations in the laboratory. If so, counting the mutations would be a good way to select workers who might be less likely to get cancer from exposure to toxins.

I found this proposal offensive, since it was already known that methylchloroform induced cancer in humans. Why not concentrate on diminishing exposure to the toxin rather than depriving workers of their jobs? The whole thing reminded me of the infamous Tuskegee experiment of the 1940s, where African American men exposed to syphilis were deliberately left untreated so as to study the course of the disease.[17] But I felt like an outsider for actually caring about our research subjects. My concerns were apparently inappropriate. Our letter to *Science*, asking whether it was ethical for scientists to collude in the exposure of workers to known carcinogens, was refused publication. On the other hand, I don't think the study was actually done, since I have never found a trace of it in the published scientific literature.[18]

My collaborator Ted Bradley and I were quite proud of the fact that we had the first information on radiation-induced mutations in real cells inside living humans (the cancer patients). We foresaw important medical applications in preventing the toxic side effects of radiotherapy treatments, and in preventing needless exposure of workers to radiation.[19] Since we knew that *Science* was interested in this topic because it had publicized the methods we were using, we submitted our paper to that journal. By return mail, the journal informed us that it would not even consider our paper for publication; they clearly thought it of less interest than the test to screen out new hires. Learned journals' usual procedure is to send proposed articles to other scientists to review, but *Science* and other top-tier journals only bother to do this with articles whose topic the editors find worthy of attention. They told us they didn't find our topic interesting enough to review. We eventually published our paper in a journal with much lower impact.[20] I was getting disillusioned with my scientific community.

At the genetics meetings, I was spending an increasing amount of time in my hotel room or swimming up and down in the tiny, overheated swimming pools. I felt alienated. I was not only taken aback by the lack of empathy toward workers but also appalled by the researchers who saw environmental mutagenesis as a competitive sport. I had blundered into the forefront of mutation research, and there were two smart young researchers from different laboratories in the U.S. who had arrived there just before me, Drs. A and B. Dr. A was helpful to me within the limits of his own interests, but Dr. B was a tougher specimen. At one large meeting, he gave a very technical talk about a procedure I was using, and I didn't understand him. During the question period, I asked for clarification, and he took up the rest of the question period with a long-winded answer, which I still didn't understand. I asked people around me, but no one else could explain it either. I went to find him after the session and asked for more details. Dr. B replied blithely, "You didn't *really* think I was going to explain how I did it? We're going to patent this!" His whirling words had been a smokescreen. He had led the whole meeting into confusion and self-doubt, including the students we were supposed to be training. When I objected, he explained to me that he didn't want to lose any advantage over Dr. A (he wasn't worried about me – younger, female, and tucked away in Canada out of range of the more opulent U.S. funding sources). Money was critical to all of us, since making human cells survive long enough to be tested for mutations required large amounts of expensive culture media. I realized that if I wanted to be in the loop and be kept up to date on how to study mutations in humans, I would have to become a satellite of either Dr. A or Dr. B and participate in their rivalry. The health of the workers and their children would become secondary.

On the other hand, I wanted to continue with the genetics research because I felt it important and we were making progress. I felt some responsibility since we were the only molecular genetics researchers I knew in Quebec who were primarily interested in workers' welfare. But our laboratory was very expensive and difficult to maintain, and the payoffs of our research for workers were going to be years away.[21] Explaining our very technical results to workers so that they could participate in orienting the research was hard and I was not particularly good at it. Being re-trained in ergonomics, a science that analyzes work in order to propose immediate concrete changes, began to appeal to me.

At that time, CINBIOSE graduate Nicole Vézina was just finishing her ergonomics doctorate in France. When Nicole came back to work at CINBIOSE again, she fascinated me with her novel approach to workers. She actually observed and analyzed the work, spending long hours figuring out how people worked, why they worked like that, and what obstacles in their work environment were dictating some unfortunate strategies.[22] We collaborated on a study of how laboratory workers protect themselves against radiation[23] and she impressed me by seeing all kinds of phenomena I had missed. She noticed that the workers dealt with radioactive substances in different ways. Some imaginative technicians visualized the contamination on their gloves and would take the gloves off before touching a telephone. Others thought of the gloves as a barrier to radiation and would keep them on all day, thereby contaminating telephones, doors, and other surfaces. Some saw putting radioactive waste in the trash as equivalent to making the radiation disappear and others were very careful with the trash.

It began to seem to me that even if I wanted to protect workers from radiation I might be better off studying ergonomics on my next sabbatical leave than continuing with my research on genetics. Ergonomics seemed to be a bridge across the empathy gap, so I went to Paris in 1990–91 to the centre where Nicole had trained, and I registered for their ergonomics practicum, a supervised intervention. My life partner Pierre came with me and we remember that year as one of the most exciting we have spent together. Pierre, a journalist, met with French journalists and writers, and I got to talk with the authors of all my favourite articles. We had dinner one evening with Christophe Dejours, a pioneer in occupational psychology,[24] and another a few nights later with Catherine Teiger, a leading thinker about ergonomics.[25] I had frequent conversations with Danièle Kergoat, a sociologist who had been the first in France to write about how physical task demands are divided up between women and men.[26] I was taught how to do computerized analysis of occupational health data by Marie-Josèphe Saurel-Cubizolles, whose group had pioneered this approach. I had never been so stimulated. And all of this was happening in historic mansions in Paris where research was housed in tiny, uncomfortable, ancient rooms, imbued with tradition and the smell of moulds. We lived in an immigrant area in the 10th arrondissement, right near the Gare de l'Est railroad station, and I biked home every evening through streets filled with exotic clothes and strange food. After supper, we would go for endless walks and bike rides by the Canal Saint-Martin, along the Seine, through the parks, up and down Montmartre and Montparnasse.

My course in ergonomics was the final stage in French training, since my previous education, research, and experience had enabled me to skip the introductory courses. At the first course meeting, the professor asked if we had any idea of the workplace where we would do our intervention. I said I was interested in women's work and in cleaners. After class, a businesslike young woman came up to me and suggested I work with her. "Anne" was the occupational health doctor for the firm that cleaned the trains at the Gare de l'Est, and many of the workers, especially the women, were having musculoskeletal problems, mostly backaches. She was worried that soon she would no longer be able to certify them as fit for work. She took me on a tour to the train station and, to my delight, the workers were out on a one-day strike and thus all lined up at the entrance. Anne, who had a contract with the employer to worry about, was nervous about my speaking to them in his absence, but I played on my status as ignorant foreigner and hung around with them for a while. They explained some of the issues in the strike and I asked them whether they would be interested in an ergonomic study. They didn't say no, although I later found out that they were suspicious of this strange woman poking around and immediately went to consult the union higher-ups.

A few days later, with Anne having taken care of all the protocol with the employer, I went to the train station for my first observations. Nina Khaled, a lively young blonde from an Algerian Berber background, wanted me to observe her. She was particularly interested in showing me her big blue pail, which to her mind was too heavy and full of useless stuff prescribed by the employer. She was annoyed because the government-owned train corporation didn't allow her to use the most efficient tools to get the cleaning done, for fear of their being too rough on the train surfaces, and encumbered her instead with a pail full of inadequate tools and chemicals that she was required to carry around all day.

Nina and I went out to the trains together and I watched her work for a couple of hours. She was assigned to clean the toilets in all suburban commuter trains and some other trains during the brief time that the trains were in the station. She ended up cleaning about 200 toilets that day. At first, like most beginning ergonomists, I couldn't see any sense to her cleaning. She sometimes started with one part of the bathroom, sometimes another, she sometimes skipped a bathroom or a part of it, and she sometimes skipped a whole train. But as the months passed and I observed her for at least a day every week, Nina's strategies emerged: the way she instantly evaluated each bathroom, deciding what absolutely had to be done as a function of the time she had to do it before the train would leave; the fact that she could only clean the bathroom if the guy who was to fill the car's water tank had gotten to it in time; the standard of cleaning that varied by type of train.

I eventually produced a pathway diagram that covered two 16" x 28" sheets, showing all the choices she made on her route, using triangles, circles, diamonds, and squares to illustrate actions, places, and alternatives. I was afraid it was too technical, but I needed to have it validated by the cleaners. I spread it out on the floor of the women's locker room at break time. Nina came in, took one look at it and immediately said, "You forgot the balls" (the task of putting little scented balls into the almost inaccessible deodorant container that was scrunched in behind the toilet). She did agree that the diagram, by and large, represented the choices and cognitive demands of her work. I was relieved that she found it obvious and gratified that she recognized my representation of her job – maybe I could be an ergonomist.

Nina's job was extremely demanding physically as well as technically. I measured the distance she travelled in a day with a pedometer: 23 km. We raced back and forth from one track to another as the trains pulled in and out of the station. She had 60–120 seconds per bathroom. Nina had to twist and bend to get into every corner and she scrubbed the toilets while on her knees. To go more quickly, she would mop the floor by pushing a soapy rag with her feet while sponging off the sink and the mirror with her hands – she called this her "dance."

This was during the first Gulf War, and, especially on weekends, trains were filled with soldiers returning to Paris on leave. The hard-drinking parties often left their traces in the toilets. The instruments in Nina's pail were often insufficient to deal with the caked-on stuff in the toilet bowls, and she would have to bring out her secret arsenal – chemicals and a scraper bought with her own money that had to be hidden from the inspectors sent by the Société nationale de chemins de fer (SNCF), the French national railway company, because of the risk that it would damage the porcelain.

Despite these unsavoury aspects, it was fun to follow Nina at work, because I had never met anyone like her. She, her sister, and two cousins had left their high school and their small French village one night because their families were about to marry them to older men from Algeria. They had taken the train to Paris and were in a café at the station discussing how to find work when an SNCF union head overheard them. He told them the cleaning company was hiring. The jobs were regular and relatively well paid, and Nina, her sister, and a cousin were immediately taken on. The three cousins were able to afford a one-room apartment in the centre of Paris and had been having a wonderful time exploring city life for the previous eight years. They invited me to dinner and showed me inexpensive ways to have fun in Paris.

Nina and her friends in the cleaning company had opinions on everything. They read the newspapers at break time, sitting on cold cement blocks outside on the railway platforms. (They were not allowed to sit on the train seats, ever.) They discussed the war, the clothes of the ministers' wives, and the behaviour of politicians. Although, unlike North Americans, the French rarely discuss their personal lives at work, my presence made it relevant for the women to talk about their aches and pains. Nina and her sister were the youngest, and they were OK, but their older colleagues had a lot of problems. "Madame Ayoub," in her forties, had chronic backache and "Madame Amin," a woman in her sixties whose back was permanently bent, hurt all over. They complained about the fact that, despite notes from their doctor (Anne) restricting them to easier tasks, they were still being assigned to cleaning toilets. There were not enough women employees, and since it was unthinkable that their male colleagues clean toilets the supervisor felt obliged to assign them to toilets from time to time.[27]

Despite our friendship, the women were never really sure they could trust me. One incident showed us they might be right. For our course, Anne and I needed photographs of the cleaners' working postures. We needed to show the bent backs and the twisting needed to access parts of the bathrooms and train cars. We wanted to demonstrate that the car designers had failed to think about how the cars would be cleaned. Anne was supposed to get permission for the photographs and I was scheduled to bring the camera to the station one Monday in February, which happened to be my birthday. I arrived straight from my birthday lunch and found that Anne had forgotten to let the workers know they were to be photographed. I nevertheless asked Mme Amin if I could photograph her as she cleaned under the train seats and she replied, "You won't get me with my rear end sticking out." I asked Mme Ayoub whether it would be OK to photograph her while she was sweeping and she replied tersely "You give me your blouse, you can take my picture in it." It then penetrated that I had been lacking in the most elementary sensitivity. My $50 birthday blouse with the gold threads could have been picked out on purpose to show Mmes Amin and Ayoub how inferior their status was and how clunky they looked in their uniforms. It was as if I had planned to humiliate them, and I had succeeded. I apologized and went home. I spent the night with gritted teeth ruminating over what I had done. Around midnight, I phoned Nicole Vézina in Montreal for comfort, but she just gently confirmed that I had been an idiot.

But after a few days, with the help of my friends, I was able to push myself back to the train station. I didn't wear the gold blouse. With proper notice, Nina was quite happy to be photographed in her new leather jacket and I got my pictures. Anne and I wrote a report analyzing the job, showing the working postures, and making a number of recommendations. We suspected that Mme Amin's back problems had something to do with the primary job she had been doing for over twenty years, which was crouching down to clean under the seats of the train cars. Remembering the classic anthropological text about Krióvrisi,[28] the village where the women's backs were all bent because their brooms were too short, we suggested an extensible brush. We thought the SNCF should buy Nina some more effective materials and lighten the weight of her pail. We even questioned the division of tasks by sex that reserved some very easy tasks for old men but offered no relief for old women. We concluded by suggesting that those who designed and built the train cars be required to ensure that their components could be easily cleaned. We submitted copies of our report to the cleaning company, to the union, and to the SNCF. Nina and her friends got their own copies and were quite happy with our recommendations. The union higher-up who had told the workers not to trust us finally met with us face to face and discussed tactics.

Anne and I got the highest grade ever given in that course, and our presentation in class was very well received. We easily published articles from this study in the academic occupational health and safety literature,[29] one of which was used as a reference for teaching graduate students in the very ergonomics program that had trained me.[30] I went back to Canada and was accepted as a full member by the Association of Canadian Ergonomists. Three years later, I received our province's highest award for interdisciplinary research.[31] My professional advancement had been given a big push by the cleaners at the Gare de l'Est.

Despite all this recognition, when I went back to the Paris train station to see Nina two years later, she greeted me by shaking her pail at me – it was the same big blue pail, loaded with the same prescribed items. It was just as heavy as it had always been and had to be carried around for the same twenty-three kilometres per day. Nothing, absolutely nothing had changed, except that Mme Amin had finally been invalided out of her job. I still had my gold blouse, and Nina still had her blue pail.

Appalled and ashamed, I called the man in charge of ergonomics at the

SNCF, a friend of one of my teachers. We met and I presented our conclusions and talked of how little money it would cost to get a lighter pail and some more effective cleaning agents and tools. He smiled and nodded but it was pretty clear from his manner that changing the SNCF's instructions to the cleaning company to accommodate a bunch of Algerian immigrants was not a high priority for his service. When I visited the Gare de l'Est on my next trip to France, I couldn't find Nina but the blue pail was still there.

It was becoming clear to me that making changes in cleaning work needed more than just publishing the "truth" about their working conditions. But I had no way into the system in France. I hoped that things would go better in Quebec. On my return, the educational service of the Confédération des syndicats nationaux (CSN) union[32] asked me to give sessions on health and safety to ten groups of hospital cleaners in Quebec based on my experiences studying cleaning. I asked them to start out by telling me about physical and mental health risks in their jobs.

Workers worried about the fact that they were exposed to disease but were never trained about how to avoid infection. They were anxious about the possibility of contamination from patients, a possibility that they had no way to evaluate because they had no access to information about the patients' health problems. Several mentioned that they had seen a patient's room posted "in isolation" after they had been cleaning it for several weeks. They might have been exposed to contagion, but patient confidentiality prevented the nursing staff from telling them what the patient's illness was. Reassurance from staff that there was no danger to them was not very helpful, since they could see that the patient had had the same illness for some time: if the patient was dangerous now, why not then? The part of the educational sessions dealing with how to protect oneself from infection from patients, given by Ana María, was by far the most popular.

The cleaners also spent some time talking about their heavy carts, poor equipment, and odiferous chemicals, but they were most eloquent about their hurt feelings. They said that on many wards, they were left out of Christmas celebrations and birthday parties. When a patient died, often the family would bring presents for the ward nurses to thank them for their attention but would forget the cleaners, even though it was they who had spent the most time in the patient's room.

Silence filled the room after recitals of a series of humiliating moments: being talked through, being insulted in front of patients, and being forgotten. I asked, "What strategies do you use to cope with this type of behaviour?" One well-dressed woman in her fifties explained: "If they don't respect me, I can respect myself." She said that she always wore stockings and high heels, even though they made her work more difficult. A younger man countered, "I dress to show them I don't have any respect for them" – in jeans and old T-shirts. Several described with pride having won skirmishes around the cleaning of human excrement, usually with nurses as the antagonists. In one such episode, described in the session in Quebec City, a nurse had spilled a bag of urine on the floor and ordered the cleaner to come back from her break to wipe it up. The anecdote ended with an apology from the nurse for her high-handed ways.

Cleaners felt acutely their position as lowest-status occupation. They told us: "We're the assholes"; "We're at the bottom"; "We're a group apart"; "We're the hospital trash." Everyone, from head nurse to nurses' aide to patients' families, had the right to criticize them. Often complaints about cleaning were not made directly to the cleaners but were voiced loudly in their presence. The nature of cleaning attracted criticism, both because cleaning interferes with other activities and because a cleaned area quickly gets dirty again. Cleaners confronting rooms with sleeping patients felt trapped: if they woke the patients while cleaning, the patients complained, but if they didn't, patients might assume that their rooms hadn't been cleaned. If they passed a room where a nurse was talking to a patient they would note the room and come back to it later, to respect the patient's privacy, but the nurse and patient could be left with the impression that the room had been skipped.

Cleaners cope with lack of respect in part by taking pride in their work. They feel competent when they can make all the surfaces shine. When asked what was most important ("What would you do first if your time was limited to an hour or so?") cleaners replied almost unanimously "*Le look*" (appearance). The cleaners therefore found strategies to make their work visible. One left the light on in the bathroom when she was done. Another always left a smell of bleach in the bathrooms. Others described strategies such as moving the curtains to one side when the rod had been cleaned, always leaving the wastebaskets in a specific place, making sure that the floor smelled of floor polish. One group of cleaners got together and marked out a space near the entrance to their hospital where they did not clean for several days, letting visible dirt accumulate so that the public and the employers would understand that their work was necessary. No dramatic rise in the prestige or power of cleaners resulted, however, and we worried about the fact that several of the strategies cleaners used to make their activities more visible involved increasing their exposure to strong chemicals that were dangerous for their health.

After the educational sessions were over, Céline Chatigny and Julie Courville, graduate students in ergonomics, helped me do a study of cleaning. We spent months observing hospital cleaners and documenting the physical and cognitive demands of their jobs.[33] We showed how hard the job was for both men and women. Our eighty-four recommendations got a much more positive reaction from the Montreal hospital than had those submitted to the French cleaning company. We were able to meet with the head of human resources and he took a number of steps based on our report. In fact, our suggestion of abolishing the division of tasks between women and men was adopted by most of the hospitals in Quebec.[34]

However, when we did a follow-up study at the same hospital twelve years later, we were disappointed. Most of our changes had disappeared. For example, one of the tasks of women in 1994–95 was emptying wastebaskets. They did this by removing a full plastic bag and replacing it with an empty one. Since the wastebaskets were all of different sizes but there was only one size bag, they had to tie a knot at the top of the bag to keep it tight around the lip of the wastebasket; otherwise it would drop into the basket and the garbage would end up splashed on the outside of the bag. After tying over a hundred knots a day, some reported hand/wrist problems. So we recommended buying more sizes of plastic bags to fit the different wastebaskets, and this was done. But when we came back in 2006, there was only one size of plastic bag, and the cleaners had gone back to tying knots. No one remembered what had happened since the head of the service had changed several times since then.

An even more important recommendation had left no trace, either. In 1995, we told the head of human resources that the cleaners said they were treated with contempt, excluded from the Christmas party, and housed in the basement of the hospital. This manager immediately organized a series of discussions between the cleaners and all the other hospital occupational groups for which he was responsible, intended to sensitize the nurses, orderlies, and nurses' aides to the needs and skills of the cleaners. But no trace of this activity remained in 2006, and the caring human resources manager was long gone. We counted, and only about a third of our recommended changes were still in effect.[35]

Why was this? What made the changes disappear? How could all trace of them vanish? A clue could be found in one of my most embarrassing moments in the 1994–95 study. At the beginning of the study, I went to the hospital to meet with the group of cleaners to present the research project and ask for their collaboration. The head of housekeeping showed me into the auditorium, and we had some discussion about how to arrange the tape recording equipment. We had been there for several minutes before he spoke to a woman who was dusting the seats in the auditorium. I realized that I had pushed past her and her cart (which was in the aisle) without saying a word and that I had been totally unaware of her presence.

While I accompanied another cleaner on her rounds, an approaching couple tossed a ball of trash into the basket she was pushing, barely missing her. I felt insulted for her, but the couple said nothing and the cleaner did not appear to notice. For me, and for others around me, the cleaners were invisible. Indeed, cleaners later explained to us that it was part of their job to be invisible. They were expected to "disappear" when there were visitors or health care personnel in the rooms on their route. The results of their labour were only visible if their work had been done badly.

This invisibility is what made it possible for the employer to make major changes in cleaners' working conditions without even noticing. One day, cleaners came to work in a Montreal hospital and found that mirrors had been installed on walls in an entrance area. A designer had suggested that the area would look bigger and nicer that way, and no one on the long list of decision-makers in the hospital bureaucracy had thought about the implications for cleaners. No one had remembered that mirrors show dirt better than most surfaces and take a long time to clean. Result: the cleaners whose route included the entrance had less time for cleaning other surfaces. Also, since it is almost impossible to keep a mirror in an entrance area looking clean during a Montreal winter, the cleaners lost a little more pride in their work.

This was not an isolated instance. Cleaners and even their supervisors were almost never consulted on the choice of flooring, wall, and furniture surface materials, yet they were blamed when the new black, rough-textured office furniture always looked dusty.

In my experience, cleaning is a refuge for all kinds of rebels attracted by the profession's relative autonomy and absence of supervision. Because cleaners move around from one area to the next, their supervisors can't stand over them as they work. Therefore, cleaning allows a lot more freedom and initiative than an assembly line factory job or even waiting on tables. But it is often taken for granted that cleaners function at a low level and have no technical expertise. One hospital sought a grant for integrating the mentally disabled and placed the young people in the cleaning service, saying "Here's some help for you." The already overworked cleaners, who were not trained in special education, were at a loss to find the time to supervise the new arrivals. They were insulted that management thought that teaching new people to clean would take no time or energy, even those who were cognitively challenged.

It is not surprising that the Ministry of Health chose to cut cleaning when it was looking for ways to save money for the health care system. During the years 1994 to 2002, the number of hospital cleaners employed by the Ministry declined by 21 per cent. Then, a funny thing happened. A wave of infectious diseases engulfed the hospitals: there were 1411 cases of *Clostridium difficile*-associated diarrhea in Quebec hospitals during a 10-week period in 2004 and 311 *C. difficile*-associated deaths in the year that followed.[36] *C. difficile*, antibiotic-resistant stapphlococci, and their cousins persuaded someone in the Ministry that cleaning might be important. Authorities saw that the lack of interest in cleaning was costing money and lives, and the workforce assigned to cleaning rose by 5 per cent over the next 2 years.

By the time some new students and I started our follow-up study in 2006, the importance of cleaning was being recognized, and the head of cleaning told us that his staff had increased. There was even a report on hospital design considerations that recommended making surfaces more accessible to cleaners and many other ways to make cleaners' jobs easier. However, although public health leaders had recognized that cleaning was key to disease prevention, they didn't see cleaners as a key population. Among the 12-person committee that produced the report, there was no cleaner, no representative of cleaners, and not even a cleaning supervisor.[37] It seems to me that someone with a practical knowledge of cleaning would have added high quality information on the choice of materials for walls and floors and would have suggested that they add a chapter on design of toilets. A cleaner would probably also have helped think of ways to solve other problems of importance to disease prevention, such as how to design patient rooms so that they could be cleaned while ensuring privacy for patients and how to design hospital traffic patterns so that corridors can be cleaned.

What would have to happen so that cleaners could be included in a committee like that one? It would not be easy, either for the cleaners or the others on the committee. The cleaners would have to be given some kind of training so as to understand disease transmission so that they could participate fully in discussions. Also, the cultural differences between cleaners and other committee members would have to be dealt with, optimally by including people with interpreting skills. But in order for the necessary resources to be allocated, the Ministry would have to be convinced that cleaners have an essential point of view. It would not be easy to drag the Ministry across the empathy gap – I had seen how hard it was to make cleaners seem important to the French railroad company.

I think it would be worthwhile. Cleaning in health care establishments is a highly technical job; it requires handling chemicals, choice of techniques and materials for cleaning different types of surfaces, a sense of organization so the cleaner can deal with the very high number of interruptions without forgetting or skipping any areas, some understanding of contagion, and a sense of tact in dealing with the people who get in the way.

However, employers and the public don't see cleaners as knowledgeable, to their loss as well as that of the cleaners. And, if cleaners are not consulted on changes that directly impact their own jobs, they are not even dreamed of when health care is being discussed. In long-term residences for the aged, where doctors and nurses pass seldom and orderlies' shifts change constantly, the cleaner is often the person who spends the most time with many lonely, sick residents. Cleaners could be trained in how to make their time in patients' rooms profitable and fun for both resident and cleaner. I have never heard of that. In fact, one cleaner, "Jeanne," told me in a soft voice about a horrible experience. She had enjoyed her job in a long-term care facility where she had worked for many years. She got to know the patients and had many special friends. But one day, her supervisor decided she would work faster if she didn't talk to the patients. Not only did the supervisor tell Jeanne to keep quiet, she chose to announce over the loudspeaker that people were forbidden to talk to Jeanne because it would interfere with her work. Jeanne was not only humiliated and alienated but also sad for "her" patients who would now have no one to talk to. They would all be victims of the empathy gap.

Chapter 3

Standing Still

I'm not going to go complain to the boss, it wouldn't change anything much. He's not going to give me a back rub, so I don't complain.

— *Clerk with sore back and legs working for a chain of gas stations in Quebec whose seats had all disappeared overnight by order of head office*

LIKE OTHER ERGONOMISTS, I SPEND A LOT OF TIME watching people work in order to understand why they have problems. A few years ago, I was watching bank tellers because their union was about to negotiate a new contract and wanted to know what changes to ask for. Managers had recently removed the seats from behind the counter, saying they wanted "to make the bank branch look more professional." Renée, a large, pleasant woman in her forties, worked standing at her wicket, so I stood behind her, scribbling on my pad, noting her movements and actions. It took only about an hour before I (then fifty-two) had an intense backache. Out of pity, the manager found a seat for me behind Renée and my backache went away. But Renée had to stand all day, with only short breaks at mid-morning and noon break. She told me she also had a backache at the end of the day. And I could see that, as the day went on, Renée spent more and more time leaning on the counter to take the pressure off her back. So I wondered, why didn't the employer take pity on her? Why didn't the sitting professor make the bank look unprofessional? Did the manager feel that Renée was being paid for standing? But so was I; in fact, I was paid a lot more. Did the manager identify more with the university professor than the bank teller? Or was there some other reason?

Whatever it was, I suspected that the distinction had something to do with respect and that upset me. I have a lot of trouble with getting special privileges based on my membership in the upper middle class. My parents both grew up very poor. My father's father was a tailor who got encephalitis when my father was twelve, probably as a result of the 1918 influenza epidemic. He could no longer work, and my grandmother supported the three of them by doing odd jobs. Much of the time she was a receptionist at a New Jersey hotel owned by her father, until the Mafia took it over in the 1920s. The extended family and scholarships sent my father to university, where he became an engineer. He graduated in 1929, just in time for the Great Depression. His first job was smoke-watcher – he had to sit on the roof of a factory and tell management if the smoke turned too dark. My mother's mother was a sole-support sewing machine operator. My mother became an artist, in the tradition of 1930s progressives like Diego Rivera, but had to earn a living doing advertising layouts, which she hated. When my parents met, they were poor. Although my father eventually worked his way up to be an executive and my mother taught art classes in the suburbs, my family was never totally at ease with membership in the middle class. Things could so easily have turned out differently. My mother, particularly, was very sensitive to issues of social class.

So when I see that workers' efforts are undervalued, that their difficulties are underestimated, that their pain is discounted, I feel concerned. I don't think Renée's back pain should have been invisible to her manager; in fact I think he should have given her back her seat. Same with grocery store clerks and sales personnel in general. I have seen these clerks sitting in Greece, France, Italy, China, Sweden, Peru, Brazil, Thailand, and Cameroon. To me, the fact that so many North American workers are forced to stand at work needlessly is the outgrowth of an empathy gap among employers, the public, and scientists.

Trying to Get Seats

It is generally recognized that it is unhealthy to stand for a long time and many jurisdictions, including Quebec, have limited prolonged standing by law. Article 11.7.1 of the Quebec *Regulation respecting industrial and commercial establishments* provides that a seat shall be supplied to workers "where the nature of the work process allows it." In 1989, citing this article, Madame Girard, a middle-aged supermarket checkout clerk who suffered from back pain, asked for a seat. Her employer refused, on the grounds that the nature of her work did not allow standing, and the case went to arbitration. The union asked my CINBIOSE colleague Nicole Vézina to be an expert witness on whether cashiers could do their work sitting. Nicole, my mentor in ergonomics research, is a self-effacing woman with a brilliant mind, a sharp eye, and total devotion to improving workers' health. She had observed cashiers' work for several years, documenting how they handled heavy weights like bottles of bleach and unwieldy cases of beer; she had already given the employers a lot of suggestions for making the job easier.[1] In the middle of that winter, Nicole wrapped up her nursing infant and, with a helpful student, drove several hundred kilometres to the hearing. She explained to the judge how easy it would be to supply a seat for Mme Girard. In 1991, the appeals tribunal found in favour of Mme Girard and her union. I was in France when the decision came down, but I heard there was quite a party at CINBIOSE.

I am writing this book twenty-two years later, and out of my window I can see the supermarket on the corner where I shop about once a week. When I enter the store, I see on the door a blue decal with the insignia of the union, and many of the checkout clerks actually wear union identification on their uniforms. I know the health and safety officer of that union fairly well; she is active and devoted to her members' health. But every one of those checkout clerks still works standing and many have told me that their backs and legs hurt, especially when they are in the smaller checkout stations where they can't move around as much. Why, in the light of Mme Girard's victory, has there been no change? Pressure to change this could have come from the Québec Workplace Health and Safety Commission, the public health authorities, the employers, or the workers themselves. Why hasn't it?

The Commission

The easiest actor to understand is the Commission. Unlike the situation in some other jurisdictions, Quebec's Commission, financed entirely by employer contributions, is both enforcer and insurer. That is, it acts to prevent workplace injury by such activities as inspecting workplaces, running campaigns for safety, and supporting safety training and research. But it also acts as an insurance company to pay out claims if workers become injured or ill because of their work. It is therefore not surprising that much of the prevention activity is concentrated on the health and safety problems that cost the Commission the most money.[2] Costs arise primarily when the workers have health problems that can be clearly linked to their workplace and cause prolonged absence from work. But conditions like prolonged standing don't cause single, easily identifiable health problems with a clear link to standing. On the contrary, they cause a myriad of different problems that emerge over time, are sometimes hard to diagnose, and can be confused with the effects of aging, overweight, or fatigue.

Prevention could also come from inspectors who visit the workplace and write an order for change, but when an inspector sees a worker standing, the sight doesn't necessarily ring the same kind of alarm bells as when s/he sees a shaky ladder or a fuming vat of chemicals. In order to recognize the intense discomfort that comes from prolonged standing, it helps a lot to have been forced to stand at work. But the majority of workers exposed to prolonged standing are young, low-income, and under-educated – not your usual labour inspector or judge.[3]

The Science

Public health specialists have not been ringing alarm bells either, because scientific studies haven't been alerting them. Scientists have not been very inspired to link prolonged standing to health problems, because the health damage from standing is not like breaking a leg at work where everyone sees a link with the workplace. Symptoms develop over time, slowly and undramatically. Distinguishing the effects of standing from normal aging is not easy. If a fifty-year-old saleswoman has chronic backache, swollen legs, or varicose veins, no one is going to spend a lot of time looking for a culprit.

On the other hand, many sales workers and cashiers I have talked to over the years make a clear link between their working posture and their aching backs and legs. Moreover, many can tell you which cash register is the worst for their backs and legs – usually the one where movement is most restricted. They have told me that their aches and pains were worst when they used to work in the cramped space behind a bar. (Of course, the kind of shoes they had to wear working in the bar didn't help any.) How come science doesn't yet know that mobility is related to this pain, if the workers know it so well?

In 1998, I went away on sabbatical to a renowned Swedish centre for ergonomics research, where knowledgeable librarians helped me search the scientific literature for the health effects of standing at work. By this time I was getting quite experienced in ergonomics research, but I found very little on standing, almost nothing relating to the checkout clerks' problems. When I tried to talk about it to the other scientists at the centre, they didn't understand why I was interested. Consistent with the Swedish tradition of healthy outdoor exercise, they were much more worried about the effects of prolonged sitting on office workers.[4] In fact, many scientists have become devoted to making workers stand more.[5] When I spoke of prolonged standing to one of the Swedish scientists, the editor-in-chief of an important professional journal, he told me he had installed a standing workstation for his office staff. He was an ardent jogger and thought people should sit as little as possible.

It is true that in Europe, workers are able to sit down more often and that some may sit too much at work. Europeans do sit at jobs in factories and services where they would stand in North America. Hotel receptionists, chicken processors, and postal workers all sat in Europe; even store clerks could often find a seat somewhere.

But back in countries where workers stood more often, such as the U.S. and Canada, there had been no large-scale epidemiological study on the effects of prolonged standing. When I decided to research this problem myself, I began to realize that lack of scientific interest was only part of the explanation. There had been an early flurry of research by ergonomists, but they quickly found it was hard to define when a worker had been exposed to prolonged standing.[6] What is standing? Is it standing if you are crouched? Bent over? Walking? Climbing? Leaning? Are professional basketball players exposed to prolonged standing at work? Would we expect the same health effects as among security guards or nightclub dancers? Probably each kind of standing has its own effects on the workers' bodies. For example, we would expect that walking, running or dancing, which involve contracting leg muscles, would stimulate blood circulation, while standing still or crouching would allow blood to pool and swell legs.

Luckily, ergonomists are skilled at observing work and writing down what they see, so theoretically we would just have to watch workers and see how they stand and for how long. But how can we distinguish in practice between standing and walking? That would seem easy, but it isn't, because no one can stand for more than a few minutes without moving either foot. So is the worker walking if she takes one step? Two? If she walks very slowly? And how can we decide how long workers have been standing? Do we have to watch them for a year? An hour?[7] A scientist would have to be really interested in prolonged standing to be willing to spend a lot of time working out the different effects of different standing postures.

I puzzled over this while in Sweden and finally asked the late Åsa Kilbom, arguably the best ergonomics researcher in the world, and one of its nicest people.[8] Åsa had done her Ph.D. research on the physiological requirements of work as a store clerk in a Stockholm department store at a time when most ergonomists were studying load handling and other more visibly physical jobs.[9] She had grounded her professional life in the constraints of unspectacular, low-status jobs, so it wasn't hard to get her interested in the checkout clerks. She thought it would be hard to study the effects of prolonged standing in Sweden, where workers who stood as part of their jobs had to be allowed to sit for several minutes every hour. She suggested we return to the department store she had studied twenty-five years earlier, since she knew they had been sensitized to the importance of ergonomics. Maybe there would be some position where workers stood for long enough that they could be studied.

I went downtown and found the enormous store. For a morning, I hung around incognita to see whether it was different from North American department stores. Not much. Strangely enough, I never saw any of the workers take time off to go sit down. In fact, I never saw any seats, and all worked standing. The middle-aged woman at the cosmetics counter struck me particularly, since she had a tiny, cramped work area and wore fancy, uncomfortable-looking shoes.

When Åsa joined me, we went together to meet the manager, who said the store policy was that all workers who stood must sit for ten minutes an hour. I didn't ask her where the seats were since I didn't want to be kicked out of the store before we could do our study. She gave us permission to observe the work, and I observed eight sales workers for a full workday each. I never found any seats. In fact I never saw them sit down outside break time, but I did see them using the same pain avoidance tactics as the bank tellers I had observed in Montreal: leaning on the counter, resting one foot on the back of the other leg, rubbing their back, elevating one foot. Five of the eight had pain in their feet, and the cosmetics worker had terrible foot pain.[10] I asked the ones who spoke English about the sitting rule, but they had never heard of it. Åsa and I wondered whether, in fact, Swedish clerks were exposed to the same conditions as North Americans. Perhaps working conditions had degenerated over the years. Despite the importance given to ergonomics in Sweden, we had had no trouble finding several groups of workers exposed to prolonged standing. We went back to the research institute and, to compare, we observed research staff of similar age. The only one who had foot pain was the one whose boss was the jogger. She was also the only one who usually worked standing.

With other colleagues and students, we eventually were able to define our terms and show that people who stood at work, especially those who weren't allowed to sit, had pain in their backs and lower legs. One of my students in Montreal, Ève Laperrière, was a waitress part time. She was eventually able to do a master's thesis on how standing postures of workers in factories, restaurants, and stores made workers' feet more sensitive to pain. Another student, Suzy Ngomo, showed that those who stood without moving had headaches, dizziness, and other symptoms of circulatory problems. Meanwhile, Nicole Vézina, who had started all this, was doing research on what would be the best way to organize work at the checkout counters. She recommended a so-called sit-stand chair, a compromise between sitting and standing.[11]

Catch-22: Funding Research on Workers' Health

My sabbatical had been paid for partly by my university and partly by the Swedish government, who wanted me to help Åsa devise a research program on women's occupational health. In exchange, the Swedes gave me access to the technical services I needed to do our study in the department store, and Åsa advised me on how to do the study. When I got back from Sweden I applied, with Åsa, for funds from the Canadian government. Since I was interested in how prolonged standing affected workers' health, I applied to the Medical Research Council (MRC). We wanted to examine physiological effects of standing under real workplace conditions, while carefully documenting working postures and workers' perceptions of pain.

Our project was rejected by the population health committee. In fact, it was rated lowest of all the projects submitted. The scientists to whom it was submitted for evaluation said things like "I have never seen a project like this before" and "This project should not be in this committee." I was surprised that they had so low an opinion of the project because our team was internationally recognized for our expertise in workplace health.

Wondering whether it was true that we had just applied to the wrong committee, I called the MRC. It was true that few projects in the population health committee were about occupational health, and most of those were statistical analyses of occupational accidents and illnesses. The woman who answered said that a decision about whether the committee was appropriate had been made before the project was officially submitted and that it was the right committee. After a half hour's discussion, we were unable to find another committee at MRC that would be a better fit. There was no place for a project that observed and listened to workers in the federal agency funding public health research.

Another place to get funding for this project would have been the provincial Institut de recherché Robert-Sauvé en santé et en sécurité du travail (IRSST), but their research projects were explicitly oriented by priorities determined by workers' compensation payments.[12] Since no one was being compensated for problems due to prolonged standing, this project would not be part of their priorities.[13] This was a kind of Catch-22 since no one could be compensated until science established the relationships between standing and health problems. The Workers' Compensation Board did support Nicole's research on sit-stand chairs, because of Article 11.7.1 (the one that said that workers should have chairs), but for technical development (better chairs), not to show links between standing and musculoskeletal problems.

We therefore rewrote our proposal as a social sciences project. I partnered with a sociologist and proposed a program about gender and occupational health, including a project about standing. We justified our interest by the fact that when men are required to stand at work they usually move around, whereas women's standing jobs are relatively immobile. This project passed peer review and was funded, but at a much lower level, since much less money is available for social sciences than for health sciences. Ironically, two years later, the MRC (renamed as the Canadian Institutes for Health Research) gave me a Senior Scientist award, for about the amount we had originally asked for from MRC, and this money allowed us to continue doing research on standing.

At about the same time as we were looking at foot pain and headaches,

Niklas Krause in California and Finn Tuchsen in Denmark were finding effects of prolonged standing on the carotid artery and on varicose veins.[14] With colleagues, we were able to demonstrate an association between prolonged constrained standing and pain in the back and legs. We found that work-related back pain is about twice as common among those who stand compared to those who usually work sitting, even after adjusting our analyses to take into account workers' age and whether they lifted weights at work.[15] But when Niklas, Finn, and I organized a session on health effects of prolonged standing at the 2006 International Congress on Work and Health in Milan, hardly any scientists showed up.

Although we still don't understand all the biology involved, it is clear that standing for a long time without being able to sit when they need to causes health problems for workers. We have a good idea that a sit-stand chair would help, and the cost is not prohibitive. But no one is interested – what's the problem?

A clue came from my conversation with the journal editor, the jogger whose secretary worked standing. I had gone to see him again because he had expertise in work postures and I hoped he would collaborate with our studies. He was a very smart man with a long history of interesting research. But during our lunch together, I was completely unable to interest him at all in the problem of standing workers. I pointed out the kitchen worker who stood behind the counter to serve the scientists in his institute's cafeteria and suggested that it would be worthwhile taking measurements of the changes in his legs over the two-hour lunch period. The editor just replied that I should tell the man to buy support stockings and good-quality shoes.[16] Desperate to find common ground, I reminded him that people who walked slowly through museums got unpleasant sensations in their legs and backs. That worked. "Ah, museum fatigue!" he said, suddenly becoming animated. He thought that would be a promising line of research; in fact, he had been thinking about examining museum fatigue.

It is not surprising that scientists, like others in their social class, can identify more readily with museum visitors than with cafeteria workers. They and their families and friends spend time in museums and have experienced "museum fatigue." But then who will do the research that will enable the cafeteria workers to have seats?

Employers and Managers

Nicole participated for ten years in a government committee on prolonged standing together with representatives from the three major supermarket chains, but they never allowed her enough time in their stores to finish testing her ideas on design of sit-stand seats. My own research on the effects of standing has gone slowly because it is hard to find employers who will even provide access to their facilities.

At one point, we became encouraged because a member of my family had an in with a major retailer. In response to her lobbying, management allowed us to observe cashiers at one store for several months. The human resources people even said they liked our report, which suggested how to re-organize the cashiers' work to make it more efficient but less boring and painful. But when we asked to do another study focused on making it possible for the cashiers to sit, again at no cost to the employer, they said this was not a priority and refused entry. We got similar responses from other employers (a hardware store, a clothing retailer, many supermarkets, a ski centre). More than one told us that if we did a study of standing, their workers might then expect to sit at work. They didn't want to raise any "unrealistic" expectations among their workers, so they didn't want us in their workplace.

When our results on the pain and suffering associated with standing were published, a local TV science program wanted to interview us. But for TV, you have to have pictures. The journalists were never able to get consent from any employers to film standing workers. Even the employers who had benefited from our recommendations, the factory where Ève had done her master's thesis, the retail stores where we had observed workers, and the local grocery stores refused. "I don't think [the factory] wants to go on television with the image that working there is bad for health," said the human resources manager. So the employer is aware that there is a health problem. In fact, during a meeting with a major retail employer about work schedules (see chapter 6), the negotiator from the employer's side explained to us that schedules were constrained by the fact that workers could not put in more than eight hours per day on the floor, "because standing is very painful. After a while your legs and feet start to hurt all the time." So these employers are aware they are causing pain – why don't they want to stop it? What is their reasoning? Is it something about social class?

Social Class

Nicole tells a not-so-funny story about a time she went to a supermarket to meet with its manager. Alone in the office, the manager explained to Nicole that a sit/stand seat was not a good idea. Cashiers *had* to stand while serving customers because it wouldn't be polite for them to sit. "When you go to someone's house, they always stand up to greet you," she said from the seat behind her desk. But she never stood up and she never offered a chair to Nicole.

It is not only store managers who are allowed to sit while receiving visitors. No one has suggested that doctors rise to greet their patients or that kings get up to receive their subjects. Standing is perceived as a gesture of courtesy offered to people of superior social status. But, as French cashier Anna Sam has pointed out,[17] this profession is so disdained that mothers bring their children to watch cashiers at work in the hope that they will be frightened into studying hard. "You see, if you don't work hard at school, you will end up a cashier like the lady," said one French customer to her daughter, in front of the cashier who was serving her.[18] In North America, their inferior social status is what makes it legitimate, even imperative, to insist that they stand.

Researchers France Tissot, Susan Stock, and I looked at the results of the 1998 Québec Health and Social Survey of 10,000 workers to see who was standing and who wasn't allowed to sit at will. We found that workers forced to stand were young and poor; 76 per cent of low-income workers stood at work. These workers were also the ones who had very little control over their work in other ways.[19] In other words those who stand occupy the lower ranks of the working class.

One would expect public health researchers to take up the cause of getting chairs for workers, but I have found that the empathy gap affects them, too. Several years ago, I was invited to California to give a keynote lecture at an international conference on women's health. The day before I was to speak, I discovered I had forgotten to bring stockings and went to a department store near my hotel. It was near the end of the afternoon. A young African-American woman processed my transaction, looking grim. I said to her, "I hope you're near the end of your shift, you must be tired." She replied that her back hurt and her legs hurt. Always subversive, I remarked, "I bet you could fit a seat behind this counter." She answered that the boss wouldn't allow it. I asked, "Wouldn't your boss be better off if the employees were happier, wouldn't they smile more at the customers?" She responded emphatically: "I don't think my boss cares whether or not I smile at work."

The next morning, I told the story about this cashier as an introduction to my talk on women's occupational health. I wanted to make the point that many women workers' problems appear trivial to decision-makers, but that women do suffer at their jobs. Indeed, I got more comments on this anecdote than on the rest of the talk, which dealt with bias in scientific research.[20] In telling the story, I had expected the audience of North American women health scientists to be sympathetic to the cashier and worried about her sore back and legs. Instead, the audience reacted as shoppers. Most of those who commented said that being served by a seated checkout clerk would make *them* uncomfortable.

It is possible that this reaction is at least partly responsible for the back and leg pain of cashiers and other standing workers. In North America, people of a certain social status see nothing incongruous in making workers suffer pain all day so that the customers won't feel emotional malaise during their shopping transactions. The scientists and health professionals who aren't interested are not mean, but they lack empathy with the cashiers.

This empathy gap has costs not only for standing workers, but also for the quality of science. There are major gaps in the understanding of physiological phenomena related to standing. We saw this during a lecture given in our biology department. A young physiologist concluded his review of studies on fatigue by saying, "You shouldn't concentrate on subjective sensations, but get objective data." Although pain and fatigue as such do not have a defined "objective" existence outside the sensations of those who feel them, some phenomena like the behaviour of oxygen in the blood are associated with feeling tired. Despite this, the speaker's presentation offered no explanation for the fatigue and pain of the workers we had met. When asked about fatigue associated with standing, he replied: "These are emotional questions."

One of the students, Vanessa Couture, who had been a sales clerk, described patiently to the speaker how she had worked in two locations. Like many others, she felt much more back and leg pain when she was confined to a small area than when she worked at the front of the store where her area was bigger. The researcher replied that this might be due to "nervous fatigue" but offered no explanation of how this would change when the area got bigger. Vanessa was disappointed that the researcher wasn't listening, especially because her master's thesis research showed a relationship between prolonged standing, discomfort, and effects on the circulatory system.[21]

Scientists do research on subjects that pique their curiosity, and they base their articles on data they find believable. Unfortunately, doctoral programs don't provide many opportunities to learn about subjects that touch workers' lives, so the speaker's curiosity hadn't been piqued. And scientists don't encounter many situations where we learn to believe workers while we are getting our training, either.

Why Workers Don't Revolt

In theory, someone who is able to work standing in the same place for a long time should be able to do the job sitting, with minimum rearrangement of the workstation. North American workers, working through their unions, have been able to reduce their exposure to a lot of toxins, lower their working hours, get pay raises, and improve their lot in general. Just about every worker I have ever spoken to about standing finds it hard and would like to sit. So why have they not mobilized in the same way against prolonged standing? Why haven't they just insisted on seats?

With two social scientists and a legal scholar, we interviewed thirty Quebec workers, all of whom worked standing. Twenty-nine of them said they suffered pain they associated with their standing posture.[22] When asked why they stood, they gave three main reasons. First, those who worked with the public said that it was important to show, by standing, that they were at the customer's service. Second, many said that their managers would think they weren't working hard if they sat. And third, it was common that their work areas weren't organized so that they could sit; there was no room for a chair or the working surface was too high. None of the three situations constitutes a real obstacle to sitting: customers all over the world have gotten used to seeing service workers sit, managers of office workers don't think they are lazy if they sit, and work areas in other countries are designed for sitting.

Only a very few of those interviewed had ever done anything at all about getting a seat. They gave several reasons for their inaction, but the bottom line seemed to be that they had too many other things to worry about. The right to a seat was just one of many legal protections they couldn't access. Many of those interviewed had trouble just getting paid for the hours they worked. Very often, they didn't have access to the legally mandated rest breaks. It was surprising how many people told us that the only clerks in their stores who got any rest or meal breaks were smokers. Managers, especially those who smoked, could understand the need for a cigarette better than the need to sit down.

These low-status workers are also subject to a large number of arbitrary actions and invasive requirements from their employers – prolonged standing is only one of them. "Corinne," a bartender, told us she was OK with her working posture: "I'm still free in the positions I can be in, it's not like the hostess job I had where I didn't have the right to put my hands in my pockets. . . . I can lean on the counter in front or put my rear end on the counters at the back."

In fact, the main priority of these cashiers, sales clerks, receptionists, and gas station attendants was getting to work enough hours to make a living wage. They needed to be offered the hours, and they needed the hours to be at a time they could work. They needed to be able to get time off when a personal emergency came up. They also wanted to work less forced overtime, fewer evenings, fewer weekends. If they were going to struggle for a change, it would most likely be a change that had to do with their work schedules.[23] And they did not want to alienate the managers who decided on their hours by asking for seats.

Their low status meant that they would not be likely to get anywhere even if they did try to get a seat. "Carole," a bank teller with leg and foot pain, described with some bitterness her futile attempts to get assigned to sitting work. She resented the fact that her supervisors sat all day: "For them it's OK, but not for us." So, every day, a large number of North American workers stand at work, suffering from back, leg, and foot pain for no good reason, in full view of the public, and there are not many protests from the customers in whose name the suffering is inflicted.

Chapter 4

The Brains of Low-Paid Workers

To be a waitress, you need all the brains you have.
— *Christine Gagnon, food server*

BEING AN ERGONOMIST MEANS SPENDING MANY HOURS in the presence of workers, seeing the workplace through their eyes. It also means shuttling back and forth between "shop floor" and offices to discuss the observations with management. It is an ideal profession for observing the empathy gap.

The branch of ergonomics that impressed and eventually recruited my colleagues and me was founded in France in the 1970s. A group of occupational health and safety researchers made themselves available to French unions.[1] These researchers would be contacted by a workplace that had a specific problem, usually a musculoskeletal problem but sometimes a problem with work organization or industrial hygiene. They would first talk with both employer representatives and workers to understand how each group experienced the problem and whether they agreed about what it was. Then they would spend hours watching a single worker or watching several people doing the same job so they could understand as much as possible about the work process. They would ask the workers questions about anything they didn't understand – why do you sit that way when assembling electronics components, why do you say that phrase to the client? Their aim was to understand the exact determinants of the workers' activity, so as to remove those that were causing the problem.

For example, if a group of sewing machine operators had pain in their shoulders, the ergonomists would watch them work and identify all the movements they made that were not good for their shoulders. Then they would work out, by observation and questioning, just why the workers were making those movements.[2] Was a shelf too high? Was a machine badly adjusted? Was the transfer of work from one person to another badly synchronized? Once the immediate causes of the risky movements were exhaustively identified, the ergonomists would attack the causes of the causes. Was the shelf too high because the worker's seat was too low? Or because the operation should have been done in another department? Was there a problem because the foreman didn't listen to the workers, or to women workers, or to immigrant workers? From a request for intervention at a specific worksite, ergonomists could analyze work organization and design that applied far more widely. Both unions and employers found that this kind of analysis gave rise to useful suggestions.[3]

My favourite ergonomist and role model, Catherine Teiger, was one of those in the founding group. In 1983, I learned of her work when the women's committee of one of the unions, the CSN, decided to hold a meeting on women's occupational health to bring together scientists who worked with community organizations. For Donna Mergler and myself, who were feeling isolated among biologists, this meeting was very important. We needed to hear that you could be a competent scientist and work with community groups, so we helped organize the meeting.

Tracking down and interesting the scientists and workers' groups from all over the world was hard, and scaring up the money to bring them to Montreal was a challenge, but we were able to recruit scientist-union pairs of presenters from Finland, France, Italy, Honduras, Nicaragua, the United States, Thailand, and South Africa, as well as hundreds of union members to listen to them. We were all there on that beautiful May morning, excited that we had made it happen and that we could hear these people's accounts.

The presentations were made jointly: the workers' representatives and the scientists had been asked to explain their part of each study. Catherine Teiger came to the meeting with a French union official, Marie-Claude Plaisantin, and they told us about a project they had done with sewing machine operators.[4] Their photographs showed a lot of women bent over sewing machines, the same job my grandmother had done, except these women were sewing industrial protective gloves rather than dresses. Marie-

Claude described how the union leaders were puzzled because women were complaining of overwork and nervous exhaustion while their jobs looked fine to everyone else, including health and safety experts. The women could sit at work and they didn't have to lift heavy weights, sweat in hot temperatures, or breathe poisons like the men the experts were used to helping out. They told Catherine that they couldn't see a problem with the job; the women ought to be able to do their simple repetitive tasks while working out what they would cook for dinner or how they would help their children learn arithmetic.

When Catherine met them, she found the women who sewed gloves in the factory in a sad state, unable to do any dreaming at work. They felt stressed and burnt out. They had typically started work in the factory at seventeen or eighteen years of age and a few years later they could no longer do the work and had to leave. Catherine examined the factory records and found there were no sewing machine operators at all older than twenty-five.

Marie-Claude and her union needed to understand what was so exhausting about sewing two halves of a glove together, so Catherine went into the factory. She saw that the process started with the cutters, all men, who carefully snipped the outlines of the gloves. They would pile up the fronts and backs of the gloves, and these halves would go to the young women to be sewed together. The women, who were paid on a piecework basis, worked very fast, producing a finished glove every forty seconds or so. Not a great deal of time, even when everything went perfectly. But Catherine observed that, for more than half the gloves, something went wrong. For example, the cutters were also very rushed. Often, the cutter's scissors would slip and the two halves of the glove would be slightly mismatched. It became part of the sewing machine operators' job to compensate for the cutters' mistakes. They had to wiggle the halves together, while sewing, in the way that would best make up for the mismatch. All within the forty seconds. Or, thread could be defective and break, or fabric could be substandard and rumple. So the women would have to start sewing the glove again, still within the forty seconds. Correcting those problems time after time while producing up to nine hundred gloves per day, in uncomfortable positions, was what had been stressing the workers. At first, when the women were young and learning the job, they withstood the pressure, but as time passed the effects of the mental and physical demands began to tell on them.

The technical demands of their job had been entirely unrecognized until Catherine came along, and no one had previously addressed the problem with the cutters. The results of this research, given to the union, eventually helped to do away with piecework pay in this industry in France. Identifying the invisible task demands on workers became an important part of the ergonomists' mandate.

Catherine Teiger eventually trained Nicole Vézina, and they both helped train me.[5] I adopted Catherine's work as my model. Over the years, we came across our own examples of workers' "hidden skills."[6] One day in 1991, I was observing the job of wrapper in a noisy industrial bakery on a day when they were baking cupcakes. I walked along the line, from the big man who poured 40-kg sacks of sugar and flour into an enormous mixing vat, to the young male college student who put the heavy trays of cupcakes in the oven and manhandled them out when they were cooked, to the older women who lined up the cooked cupcakes for frosting and packing. I walked along the line until I reached the women who wrapped the frosted cupcakes. One wrapper, a woman of about 35, took cupcakes two by two from a moving line of cupcakes and placed them on a cardboard rectangle. Then she wrapped them in cellophane and put the wrapped cupcakes back on the conveyor belt to be put in boxes by the women further down the line. Finally, men would come and take the boxes away to be put in trucks.

The wrappers' job looked so easy to me and so much like what I did with leftovers in my own kitchen that I decided to use one of Catherine's favourite questions to find out if they had hidden skills. "How long did it take you to learn this job?" "A couple of days" was the answer. "And how long before you could do it well?" "A few weeks" was the first response. And, in fact, it had only taken her a few weeks to figure out how she had to sit and place her hands, in what order to hold the cupcakes and get the cellophane, and exactly how to place the cupcakes on the cardboard. But when I asked for details, I found out that sometimes she had to fix the machine that dispensed the wrapping material, or make up for bad lots of cellophane or cardboard, or place the cupcakes so as to hide any defects, or dispose of cupcakes that were too badly damaged, or help out co-workers who made mistakes, or anticipate problems from other departments. We finally arrived at the figure of two years before she had become proficient. In fact, the job was nothing like wrapping leftover chicken in my kitchen. We counted her output as 1320 wrapped packages an hour, or about 2.7

seconds per package. It takes me about 15 seconds to wrap a piece of chicken in plastic wrap and be ready to wrap another one, if everything is ready to hand.

When an employer underestimates the skills and exertion required by a job, the workers can have serious problems. Nicole Vézina tells a sad story about a boot factory that decided to modernize its production.[7] People in the plant management had heard about the so-called Toyota method that had revolutionized some assembly lines. Instead of dividing the work up so that each worker did one small segment of producing a car or a boot, now groups of workers or "modules" would work together. Each worker would rotate through the different stages of producing the product. The theory was enticing: work would be less boring, a work collective would develop and become creative, each worker's movements would be less repetitive because tasks would be diversified. Workers would no longer be paid at individual piecework rates, a pay method that scientists associate with health problems due to stress.[8] On the contrary, the whole group would be collectively responsible for the production quota, giving people a buffer against their "bad" days.

But when Nicole went to observe the work, she found catastrophe. Since workers now had to work successively at several machines, not just "their" machines, their seats had been taken away to make room for moving around. So now they stood all day and had a lot more leg pain than before. Also, each worker was now responsible not only for "her own" job, which she had been doing for ten to twenty years, but for "other people's" jobs she knew nothing about. The change in work organization had been introduced rapidly, with only a few days allowed for learning the new jobs before the group quota requirements were brought in. Each group's total production now determined every member's salary.

Workers had two choices. They could follow the new organization and work at several machines in rapid succession. But when "Annie" worked at "Rose-Marie's" former machine, Rose-Marie would get annoyed. Why was Annie holding the boot wrong, taking up time and lowering the group production that paid Rose-Marie's salary? Why should Rose-Marie have to take time from her own production to help Annie finish in time? But if she didn't help Annie, unfinished boots would pile up next to Annie and no one would be able to produce anything. Also, what happened when Annie got sick or just tired? Or when Rose-Marie had her period and didn't feel able to stand all day? Formerly, each worker could decide for herself whether to stay home and lose pay or go to work and feel awful. Now, if she came to work she faced pressure to keep up her production so that her colleagues wouldn't lose out. If she stayed at home, on the other hand, she would be replaced by an inexperienced worker who would lower everyone's pay rate as well. The combination of rush, reproaches, and resentment became toxic.

Eventually, workers were tempted to go back to doing their former jobs where they could work faster. For the group, it was obviously much more advantageous for each person to do the job for which she had the most experience. By the time Nicole arrived, many of the work collectives had surreptitiously gone back to their initial task assignments, but with more pain because of the standing.

With the help of the bootmakers, Nicole was able to redesign the production process; her team made seventy-eight recommendations, half of which were accepted.[9] By improving tools and methods, Nicole's team found ways to ease the job. They recommended a more extensive training period, making use of the acquired skills of the experienced workers. Each worker could train the others in how to do "her" job. They also arrived at a better design of the workspace and work process; although the workers still had to work standing, they moved around more.

If an assembly line requires hidden technical abilities, despite the fact that the workers are doing more or less the same task over and over (albeit with raw materials that can vary widely), what about service work, where they must adapt to an ever-changing clientele? Most people don't think of waiters and waitresses as having an intellectually demanding job but, as I mentioned at the beginning of the book, being a "counter girl" in a fast food restaurant in the 1960s was full of cognitive challenges. Dealing with dozens of clients who were on very short lunch or rest breaks also had its emotional difficulties, requiring us to prevent or deal with complaints about the food and the speed of service.

This is why, in 2005, the restaurant workers' union suggested we do a study. They wanted us to show how hard their work was, to get the employer and the public to recognize the physical, mental, and emotional job demands. But more than anything, they wanted scientists to give the public a simple message: "Tell them that we're intelligent."

Ève Laperrière, a doctoral student in ergonomics who had paid for her undergraduate studies by waiting on tables, was delighted to respond. For her master's degree, she had done a small study on foot pain and was eager to go on. The Québec Institute for Research on Occupational Health and Safety (IRSST) furnished statistics showing that manual workers in restaurants had a lot of work accidents, and that they reported a number of musculoskeletal problems. Ève had reviewed the scientific and medical literature and found very little had been published about food servers' health and safety – only three articles on their musculoskeletal problems compared to dozens on office workers' problems.[10] A few articles mentioned that food servers were especially exposed to tobacco smoke at work, but since smoking in restaurants had been outlawed for several years in Quebec they weren't too relevant to our situation.

The idea of an ergonomist observing in the workplace is scary to a lot of employers. They are afraid that someone will tell the authorities that their jobs are unsafe, so the union sent Ève first to the only place that said yes, an upscale restaurant where all the food servers were men. The pace was slower than Ève had known during her time as a waitress in a less elaborate establishment, but she was impressed by the weight of the dishes. She found that, even empty, the dishes weighed 7.5–30.6 kg per load carried. Yes, the men said, the owner had bought beautiful new dishes that were much heavier than the old ones and their arms, shoulders, and backs were suffering. It was the same phenomenon we had seen with the cleaners – their supervisors, following their aesthetic impulses, had with one thoughtless act worsened the waiters' everyday working conditions. The men also had to walk around a fair amount; Ève clocked them at about eight km per evening. Again, details of the workspace and task assignments had not been designed to minimize steps – the restaurant was on two floors, and many tables were far from the kitchen. Ève made a number of suggestions for change.

Armed with these preliminary results and lots of charm, Ève was able to negotiate her way into observing work at two family-style restaurants that employed both men and women. The negotiation was difficult because the employer's insurance company objected. When we met with the insurance agent and the employer, the latter was in favour of the study because he thought he might learn how to keep his workers from being injured. But the insurance agent was afraid Ève would discover something that would help injured food servers get compensated. He kept asking what grid we were using to evaluate the job – he seemed to be particularly afraid of one grid used by the National Institutes of Occupational Safety and Health in the U.S. Ève kept explaining that we didn't use grids and weren't intending to do a job evaluation but just understand the work. Finally she was able to get in and observe.

She found that the food servers in both restaurants had developed many useful strategies. Since they had to cover about twenty km per shift, much more than in the fancier, slower-paced restaurant, they had to save steps. For economic reasons, all the floor space was used for clients' tables and there was no place to put trays, so food servers balanced up to three plates on each arm in all sorts of ingenious ways. They would also synchronize operations so as to maximize the number of dishes they could transport at once. For instance, they would precede an excursion to collect dirty dishes with a check to see whether any clients needed something from the kitchen that they could pack out on the way back. They had dreamed up ways to carry coffee urns and creamers at the same time. They were aware that they were overloading their bodies, but "I want so much to bring it all at once." They did express some fear of the long-term effects: "We're going to end up all crooked."

The physical aspects of the job were taxing. Carrying weights, rushing back and forth, prolonged standing. And there was no place food servers could sit, even if they had the time. They were discouraged from sitting at a restaurant table because customers waiting for service might be offended. But it was the cognitive aspects of the job that were the most demanding. The servers had to handle ordering, serving, dealing with the kitchen and billing, with their eyes roving over their customers to see if any needs arose. They had to coax orders out of undecided customers and calm those who insisted on ordering dishes not on the menu. They had to remember who had ordered what and any special details about the order. They had to retain details from their initial interactions with customers in case they were alluded to later. And timing was a major problem. They had to time the dessert order so the wait wouldn't be too long between dinner and dessert, and they had to bring the bill neither too soon (customer feels rushed) nor too late (customer gets annoyed and may even leave without paying). "When you stop working, your head is still buzzing," said a waitress.[11]

Experienced food servers were able to anticipate problems and delays.

They could figure out what an inarticulate customer was asking for and recognize signals of growing exasperation. They knew the exact time to order dessert so that it would arrive when the client was ready, how to fill the coffee cup while holding it away from the saucer so there wouldn't be any coffee drips in the saucer, how to avoid eye contact with impatient clients while getting to their side as quickly as possible. But beginners had to learn to cue themselves that it was the gentleman in the red shirt who had asked for the ketchup and the one with the blue tie who wanted the water, that the next things to do should be water/bill/condiments/coffee and tea . . . water/bill/condiments/coffee and tea . . . water/bill/condiments/coffee and tea. . . .

And then there were the emotional challenges: cheering up depressed clients, coping with the whims of the kitchen staff on whom the success of the orders depended, just being eternally pleasant. Add the more serious incidents that happened from time to time: the impatient client who arrived at the busiest time, insisted on immediate service because he was in a rush, and then sat reading his newspaper for an hour after eating, occupying the place of other potential clients. The imperious client who started his order by barking: "Hey, heel!" and explained to the astonished server that he was a dog trainer and used to obedience. The gentle client who warmly thanked the server for a great job and put a $10 tip in her hand, only to watch his wife withdraw it, replacing it with a $5 bill.

The women often talked about being asked for "a nice big breast" or "I'll have your thigh" (of chicken) or being the subject of more direct remarks about their bodies: "It must be fun having those little round things [gesture at server's breasts] in your hands." But Ève was surprised to hear that sexual harassment and aggression were a common problem for men too. "No one likes having their ass grabbed, not guys either!" complained one young man.

In fact, the emotional aspects of the job added to the cognitive work. Dealing with sexual remarks was, in fact, another unrecognized skill. The food servers learned to balance the clients on the fine line between sufficient warmth to keep them happy and enough distance to keep them polite. But the fact that in North America a substantial part of the food servers' income came from tips (more than half, in the restaurants Ève studied) made that balance extremely hard to find. They were reluctant to run the risk of offending even a rude or insulting client. The client had the power, not only of complaining to the supervisor, but also of controlling the level of pay.

I learned more about tips when I was asked to go to Ontario to consult with a union-oriented clinic that was concerned about the health of casino workers. The union president took me around and gave me a worker's eye view. I saw the upside and the downside of working in a casino. When I walked into the place for the first time, I was overcome by the noise, tobacco smoke, flashing lights, and confined spaces. At the age of fifty-four, I just wanted to go take a walk outside. But the young people who were receptionists, cashiers, food servers, and security guards said it was a great place to work. They didn't mind the smoke and noise, they found it exciting. They said they were as thrilled as the clients when loud ringing and flashing lights announced a big win. They loved recognizing important local figures and watching people rejoice in their luck. On the other hand, the older workers I met were less enthusiastic about the rush and less excited by the bigwigs; they were there for the relatively high rates of pay and the big tips from high-rolling clients.

And then there was the intergenerational kerfuffle about uniforms. The casino had recently outfitted the waitresses with plenty of décolletage, and some of the women found that their tips had gone up as the necklines went down. But others, particularly older women, found it embarrassing and uncomfortable to wear the uniforms. When I looked up the scientific literature, I found that the youngsters were right: younger, blond waitresses with larger breasts do get larger tips than their older, brunette, slighter colleagues.[12] The union was trying hard to find a compromise that would allow people to be comfortable with their uniforms.

I was uneasy with the idea that food servers be compensated for the beauty of their breasts. It struck me that the problem was not the uniforms – it was the tipping system itself. In fact, tips turn a salaried job into a piecework job, by constituting a bonus for faster "production." Why shouldn't food servers be paid a flat hourly rate like everyone else in service professions? But when I suggested questioning the system, almost all the workers were energetically opposed to changing it. I had to remember how pleased I had been, at sixteen, with the first twenty-five cents I found on my tray. It was like getting a surprise birthday present. Since I was paid a dollar an hour (minus, of course, the uniform-cleaning costs), twenty-five cents was a considerable amount of money. For both the casino workers and the food servers, tips were a vital part of their income. Also, in jobs where their ingenuity, efficiency, and knowledge were often underestimated, they were comforted by the recognition symbolized by the tip. They hoped that their improving skills would be rewarded by extra money. Unfortunately, in practice, the food servers said they saw little or no relation between the trouble they took and the tips they got.[13]

Today, in Canada, tipping in cash is still reserved for those service occupations where people are not paid very much. The intent is to give an incentive to people who really need the money, giving the customer leverage to get good service. We don't tip doctors, lawyers, or accountants in cash – they would be insulted and probably not influenced by a few dollars more or less. And they already charge us what they think they're worth – a lot. Tipping is a gesture from high-status to low-status people. And customers, no matter what their social class, like the idea of being able to reward those who give them good service, just as employers like to pay employees on piecework so as to favour quicker production. Even if tips don't really improve service, they do serve to accentuate the social gap between food servers and customers.

In order to decide whether social legislation should replace tipping with higher pay for all, it would be good if health and safety experts could tell service workers and their unions whether tipping is associated with poorer health. Given the number of college students who work part time as food servers, I would have expected to find academic studies of the effects of tipping in the health literature. But when I searched the literature for studies relating to the occupational health of food servers, I found few more than Ève had a few years earlier.[14] Almost none of the studies targeted food servers' health risks, although a couple of dozen picked them up as high-risk groups in more general (population-based[15]) studies of tobacco-related illnesses (cancers, respiratory illnesses, heart disease, and adverse pregnancy outcome) and workplace injuries and homicides. There was very little in the English-language literature about the cognitive, psychological, or emotional demands of food service, although there was one article by a French ergonomist on the cognitive demands and one on bullying in the restaurant workplace.[16]

Overall, the image of restaurant food service in most of the scientific literature showed me that many health researchers see food service more as a source of risk for the public than as an occupation with health risks for the workers. For example, one article entitled "Foodservice Employees Benefit from Interventions Targeting Barriers to Food Safety" dealt only with the benefit to consumers in training restaurant workers in food safety.[17] It was stated that "the intervention was necessary to improve overall behavioural compliance and handwashing compliance." I was unable to find any "benefit" to food service employees in this article. I was reminded of my experience telling the public health meeting about cashiers (see previous chapter). When studying food service employees, do scientists often react as customers rather than as public health researchers? Do students forget their experiences as wait staff once they get their Ph.D.'s?

I don't think Ève is going to forget her days waiting on tables, but I do think that everything surrounding this job tells the customer that s/he is of higher status and deserves more consideration than a food server. An important part of the restaurant experience seems to be the sensation of being served, the illusion of privilege. But if the customer is king, what does that make the food server? In fact, food servers told us they concealed their profession from family members and acquaintances because of its low prestige. A young waitress explained how a young man broke off their online courtship when he found out where she worked. She also told us how another customer gave her compliments, followed by, "Too bad you're just a waitress." Even long-time wait staff think of themselves as temporary. Which makes them that much more reluctant to rock the boat by asking for changes in their working conditions.

I am afraid that most employers, clients, and scientists are separated by an empathy gap from understanding the cognitive and emotional accomplishments of this group and the difficulties they face. On top of this, when the servers are female, there is an added layer of distance and perhaps uneasiness that derives from the ambiguous sexual associations of this profession. The image of a woman serving a meal to a man may make him forget her professional role, especially if she leans over him with a generous décolletage.

The practice of tipping may not in itself cause food staff to be seen as sex objects, but some patrons, male and female, feel entitled by tipping to act out their sexual fantasies. They can be especially encouraged by the ambiance cultivated by some employers, whose wait staff are explicitly presented as sexual objects. A few dollars extra allow some clients to feel OK about asking for emotional support, flirting, and even going further and pinching buttocks and stroking shoulders, in situations where it may be hard for the staff to object. And the empathy gap prevents customers and employers from seeing, or caring, how uncomfortable they are making the wait staff.

At the time of writing, we don't know what effect Ève's study will have. She interrupted her Ph.D. program to have two children and has only just finished her thesis. As often happens, the union personnel who initiated the study have moved on, and the active and dynamic food server who helped us the most has left her job for another, more prestigious profession in management. One of the problems with scientific research done in collaboration with the workplace is that the workplace is a moving target – by the time the scientists have done their analyses they may no longer be relevant.

Some of the problems of food servers can be alleviated by standard ergonomic approaches such as attention to the weight of objects and the geography of spaces, providing lifting aids, and using workers' expertise to train novices. But many of the difficulties arise from the empathy gap between food server and employer or customer. It is hard to think of a solution for the fact that an employer or customer doesn't care if the server goes home and cries.

Chapter 5

Invisible Teamwork

I'm going to be all alone on Monday!

— *Bank teller looking at the work schedule for the following week*

Ana María Seifert has worked with me since 1978, when she was an undergraduate, and I really appreciate her scientific and practical talents. But every so often it feels like a mixed blessing. Like the time, a few years after our study of the hospital technicians, when my doctor ordered my annual blood/urine samples. I went to the hospital laboratory, where the personnel who remembered me from the study greeted me warmly. They crowded around and told me about the latest developments in their service. But something was not quite right. They kept looking behind me. Finally, one of them got to the point: "Where's Ana María?" They all wanted to know how she was doing and when she was coming back to visit. This happens a lot. The bank tellers, the teachers, and the hospital workers are all happy to see me when we meet again, but they all keep looking behind me for Ana María.

It is easy to understand. Ana María is warm and endlessly interested in people. She is fascinated by listening to them, and picks up on what is important to them. It was she, for example, who pointed out to me how the bank tellers put pictures of their children and their children's drawings at their wicket. She explained that people who see hundreds of clients a day to whom they represent an often-hostile institution need to remind themselves that they are human beings. And she made me understand that it was critical that tellers and others have a permanently assigned space, not be switched from one wicket to another, so they could mark their places with personal items.

Her study of hotel workers was also done in her own style. When she went to the hotel for the first time, the room cleaners were suspicious. They were from many countries, and national groups banded together in competition with others for more work hours and easier cleaning routes. Ana María had been born in Bolivia and some from other origins were concerned lest she take sides against them. They were reluctant to have her observe their work, fearful that she would report the shortcuts they took and their other time-saving tricks to the management or to snitch-minded colleagues.

Ana María finally adopted a technique that was absolutely illegitimate from a scientific point of view: she helped clean the rooms. Instead of observing, she pitched in. Although she only helped when a worker was having a problem or was not feeling well, the fact that she helped got her accepted. She says that the important thing wasn't that she made the job easier, it was the fact that she accepted doing what they did. She asked for advice on how to do the job and she listened to the answers. She developed friendships with several of the room cleaners, and still sees some socially.

At the same time, her intimate knowledge of their work enabled her to make lots of suggestions for improvement. For example, she pointed out to management that it took longer to clean a room if a new guest was coming in than if the same people were staying on, because the room had to be cleaned to a higher standard and more things had to be renewed, washed and changed.[1] It took longer to clean rooms on different floors than rooms on the same floor because of the time spent running back and forth. Ana María suggested taking such variations into account when assigning workloads. She also suggested that management think twice about such practices as offering clients with young children complimentary jigsaw puzzles with dozens of tiny pieces, or multiplying the number of little bottles of creams and lotions that lined the bathroom sink. And she noticed the complicated relationship between the room cleaners and the young men assigned to the hotel floors. The women were dependent on fresh linen to be supplied by the floor guys, and their relationship with the men could determine how long the room cleaner would have to wait. Ana María's recommendations eventually led to increasing the ratio of workers per room in Quebec hotels, countering a worldwide tendency to work intensification.

Because she is such a good listener, one of Ana María's most important intellectual contributions has been her understanding of teamwork. Even during our first study in the phosphate refinery, she was quick to see how the workers interrelated and how they shared and organized work. In the hotels twenty years later, she showed us how team members provided buffering against workload variations. When someone was confronted with a surprisingly dirty room and fell behind, her colleagues would help. When a cleaner was feeling ill or tired, her friend would come into the room and start making beds.

This angle is important to unions as well as employers, and most of the research Ana María and I have done together has come from the *l'Invisible qui fait mal* (*The Invisible that Hurts*) union-university partnership, an outgrowth of the research centre CINBIOSE that Donna Mergler and I founded in 1990. The partnership officially began in 1992 after I got a phone call from Sylvie de Grosbois, at that time a co-ordinator at the Service aux collectivités (now head of that service). She had heard that a new grant program was starting where community organizations could get research money. The funds would come from the Québec Ministry of Health, whose interest was focused on getting medical researchers to work in areas of interest to hospitals and health care providers, but Sylvie thought that the unions could get recognized as health care delivery organizations because of their work in promoting occupational health and safety. Community organizations even served on the committees that decided who would be funded. We already collaborated with two union confederations, the CSN and the FTQ,[2] so Sylvie suggested that we apply with them and with another union, the CEQ.[3] The rules allowed for an equal number of community representatives and university researchers, so we collected the heads of the women's committees and the health and safety committees of the three unions, as well as six researchers in biology, ergonomics, sociology, and law.

The first meetings were noisy and disorganized as we tried to agree on a strategy and content. Sylvie and I tried to get the union people to anticipate their research needs, not easy at first for people whose daily work involved action rather than theory. Also, even within the same union, health and safety staff, mostly men, were not used to working with women's issues staff, all women. The women's issues people didn't know the health and safety legislation and were unfamiliar with occupational risks. The health and safety guys were used to dealing with work accidents, not invisible dangers, and they were not always feminists.

Some of our collaborators were tough cookies. Carole Gingras, one of the union people, was particularly critical. A perfectionist, she started each meeting by saying, "I'm really disappointed in this version." She went over the proposal time after time, correcting the point of view, the proposed schedule, the language, and the science. We would then try again to make the proposal reflect the needs of all the unions. We sweated to articulate exactly what the unions wanted to know and how we were going to find it out. After hours of discussion, we realized that Carole, like the others, was in fact totally committed to the success of the project.

Once the union representatives were satisfied, we were all proud. Each union proposed a project in an area that, to them, typified jobs usually held by women. The FTQ wanted us to study bank tellers, the CSQ, primary school teachers, and the CSN, office workers. They wanted the ergonomists to work together to get the dangers in these jobs recognized. Also, all three unions wanted the law researchers to attack the difficulties people experienced when they tried to get compensation for health damage in jobs that didn't have visible, dramatic dangers like falling off a construction site or dying in an explosion. Hence the project name *l'Invisible qui fait mal.*

For the first time in my professional life I had written a grant application proposing exactly the research I wanted to do – directed toward improving real working conditions. We all felt good about the application and we were right. In 1993, we got a call from Sylvie: we had hit the jackpot: $361,773 for two years. The unions liked working with us and came up with new ideas for projects, and we were stimulated to think about areas of ergonomics that were new to a lot of people. And for the next sixteen years we got an average of about $200,000 per year from this source; I even got a special fellowship to take two years off to write books describing our research results. One of them, *Integrating Gender in Ergonomic Analysis*, was translated into six languages and used to train labour inspectors in Greece and Portugal.[4] We were asked to write a fact sheet and a booklet for the United Nations on gender and occupational health.[5] The team's research on law, led by Katherine Lippel, won her a research chair and many honours, as well as leading to new legislation and improved procedures for training health and safety personnel.

This program funded most of the research discussed in this book, and it was here that we learned all about the hidden teamwork in many workplaces. The first study we did with the FTQ union confederation came at the request of their bank workers' union. The job of bank teller was being rapidly transformed, and not all the changes were good. Early in 1995, we sat down with a group of tellers to hear about how they saw their problems. First, as we might have expected, they described their fear of robberies, since the robbers usually showed up at a teller's window. They especially feared "loud" robberies where the robber yelled or threatened them, but they were affected by all robberies. One teller had been present at forty-two robberies, got too nervous and transferred out of the teller job.

Effects on nerves went beyond the actual robbery. While I was observing "Danielle," a teller in a branch where there had been a robbery the day before, suddenly she gasped and ducked down behind the counter. Danielle told me later that "the way [the client] put his hands on the counter" flashed her back to a movement made by the robber the previous day, when he was slipping her a note to demand money. I asked whether she had been given any counselling after her experience, and she said that since the attempt had been unsuccessful and the robber had been arrested, she didn't need counselling. But clearly she was still panicky. Our study of tellers showed that only 8 per cent had taken time off after a robbery.[6]

One teller described a nightmare experience where she was forced to lie on the floor and the robber walked on her pregnant belly. She said she was still affected, years later. The bank had offered her counselling by a professional agency but she and others hesitated to use it because they were afraid that the bank would use the content of their agency interviews against them in workers' compensation proceedings. They didn't trust the agency to keep their secrets, since the employer was paying the bill. But they did trust and rely on each other to help keep the stress down. Danielle explained to me that she reacted less to scary-looking clients when her usual colleague was next to her rather than me. "I know you won't recognize them, so I can't do the same things." When she was working beside another experienced teller, Danielle could count on her colleague to be alert, to recognize signs of trouble, and to take action if necessary. But because of the way schedules were set up, she sometimes had less confidence in the unfamiliar teller by her side.

Aside from robberies, there were other sources of stress. Tellers described being ill at ease with their new role in marketing bank services. They had been accustomed to an accounting clerk function, where they took in and doled out money, but cash machines had made most of that part of their job unnecessary. Now, they were being transformed into vendors. They had to sell credit cards, mortgages, and car loans. Although some were reluctant to change, several had discovered a talent for selling and enjoyed the opportunity for more exchanges with bank customers. But, they said, the employer didn't make it easy. Some supervisors insisted that the daily sales quotas be met even on the first day of the month when their clients were largely welfare recipients and old age pensioners cashing their cheques. Tellers were embarrassed and upset at having to sell credit cards to these people, when they knew that credit cards might not be the best product for them. One teller was further distressed by having to put several days' "hold" on welfare cheques, even when there was no doubt that the government cheques would be honoured. The supervisors didn't have to listen to the clients' pleading for a few dollars for a child's birthday present.

Ana María went to the bank branches and spent a lot of time observing tellers. That was how she found out about a "star" system that was making the tellers especially unhappy: selling a credit card, a mortgage, or other product earned stars on a chart. The accumulated stars could be cashed in for trips or other rewards. Some tellers went all out in the race for stars. One teller had even been reprimanded for "stealing" stars by entering other tellers' sales in her own records. Most of the group were ill at ease with the spirit of competition. In fact, one personable teller who had done very well had refused to go to a dinner in a nice restaurant held for the most successful marketers in the city. She said that her sales were a team product and she resented the bank singling her out for praise. She didn't want to feel as though she was profiting from her colleagues without their getting any advantage.

While some tellers were more comfortable pushing products than others, it was true that sales were a team effort. A client would come to one teller's wicket to pay his mortgage and the teller would suggest a new credit card. The client might not accept immediately, but only after thinking it over. When he returned to the bank after deciding to go for it, chances were he would sign up for it with another teller. Who should get the star?

Employers see nothing wrong with incentive schemes like this, which are akin to the tipping discussed in the last chapter. But ergonomic analysis has taught us to look at how teams function. One phenomenon that bank clients have probably seen is the swarm; if a customer asks a teller to do something unusual (deposit foreign currency, send money to another branch, buy travellers' cheques), a group of tellers quickly forms and works out the problem together. At the bank where Ana María and I did our observations, the swarm was caused by overly complex procedures. At that time (the mid-1990s), each bank branch had a loose-leaf notebook where all the procedures for each operation were printed out. Every week, new pages arrived, and each teller had to initial them to show that she had read them and could be held responsible for knowing what they said. But this was an impossible task, especially since more and more part-time tellers were being hired. So the tellers would specialize informally. One would know all about mortgages, another about foreign currencies, and another about interbank transfers. The more senior a teller was, the more procedures she had mastered, and the more time she spent away from her wicket helping other tellers. (The system still seems to work. Recently, I needed to cash a U.S. cheque; even though the teller now had to look up procedures in the computer rather than in a notebook, the same swarm formed.) So the individual teller was dependent on the team for her success, and giving stars was actually weakening the team and hurting the bank.

Too often, the employers don't know about the teamwork that workmates have come up with, and they may even unknowingly interfere with it. Part of this is because some teamwork is illegitimate. The bank tellers who had to sign the notebook every week couldn't tell the manager that they had distributed responsibility for knowing the procedures among themselves. Similarly, when we had studied municipal workers, we heard of a cleaning team that worked at night when supervisors were absent, covering many nights for a team mate whose wife was dying of cancer. The employer never heard about that, either.

The bank branch manager may not have known what a bad idea it was to schedule six very recent hires and only one senior worker for the Monday schedule (see the quotation at the beginning of this chapter). But the senior worker immediately saw, looking at the schedule, that she was going to be besieged and anxious. In fact, one experienced bank teller we observed was interrupted more than fifteen times per hour when working with less experienced colleagues.[7]

When workers are constantly shuffled around, teamwork can't develop. About 40 per cent of women workers and 34 per cent of men workers in Canada are now hired in "atypical" work, that is, work done outside the "usual" Monday to Friday 8 to 5 time slots and/or outside the framework of permanent employment contracts.[8] The growing number of "atypical" workers is affecting all workers' ability to collaborate in teams, because even regular workers with permanent contracts must work with agency hires and casual workers.

We studied informal teamwork more directly when we looked at attendants in a nursing home, to see how they handled physically demanding tasks like lifting patients.[9] We were interested because attendants had the highest rate of compensated work accidents of any health care profession in Quebec – every year, twenty per cent of attendants had a work accident. We did sixty-one hours of observations of a total of thirty attendants.

During the sixty-one hours, the attendants communicated with each other about once every seven minutes. "OK, I'll go get a chair." "Do you need help?" Just under half of the physically demanding tasks were done by attendants working together. In fact, teamwork was an important part of how the attendants protected themselves from injury, and lifting patients in co-operation was recommended by the health and safety courses they had to take. The harder the task, the more likely it was to be shared. An attendant who had finished with her patients would often offer to help another. On the other hand, if attendants got too busy, they were unable to take the time to organize sharing.

Like the bank tellers, attendants were very aware of each other's movements, even when they couldn't see them. I heard attendants remark on the progress of their colleagues, since they could hear how far they had gotten in their workload. "She must be having trouble with Madame X," I would hear. Once, I was watching an attendant dress a patient who was in pain when another attendant came thundering unexpectedly into the room at top speed. From his position down the hall, he had heard the patient yell and was afraid his colleague had been attacked.

The employer was not taking advantage of the fact that attendants helped each other. Usually, attendants were assigned a block of contiguous two-patient rooms while their colleagues had other blocks of rooms. But Julie Lavoie, an ergonomics student, found that lifting tasks could be shared three times as often when two attendants were assigned to different patients in the same room, compared to when the two were assigned to separate rooms.

It also helps when people working together can get to know each other. An ambulance driver told me that when women were first hired the men became rapidly convinced that they were incompetent, because male-female teams kept dropping the patients. In fact, the problem was one of teamwork. When the men tried to pick up a heavy patient with an unfamiliar helper, they had trouble synchronizing their movements. They would have to agree on signals for when to lift and figure out ways to compensate for height and strength differences. When a small person and a large person lift a patient to waist height, whose waist will it be? If the weaker one lifts in two movements (ground to knees . . . knees to waist) and the other lifts in one movement, the patient (or the paramedic) may well end up on the ground. It took time for men and women to learn to work together, especially since there were so few women. This is why stable teams where the members can get used to each other's ways of doing the job are really important.

Atypical scheduling is particularly bad for teams. It used to be that people had regular jobs at regular times. Even shift work was regular, in that the worker would do so many days on the day shift, then so many on the afternoon shift, then go to the night shift. Such and such a shift slot "belonged" to a worker and the workers knew their co-workers, supervisors and surroundings as well as the way things usually went on that shift.

Now it is more and more common, in many workplaces, to build the work schedule around calculations of workers needed per hour or even per fifteen-minute time segment. As we heard one storeowner say to a management trainee, "Your schedule goes with your sales, that's all." Whether the "product" is patient care, retail sales, or telephone service, computer programs now enable employers to build the schedule around anticipated task volume. Hospitals calculate how many patients they expect to be on X ward, and the number of minutes' care each typical patient on X ward will need at each time. Stores calculate how many items will be sold in Y department on Tuesday during every fifteen-minute interval, based on holiday dates, ball games, product specials, and weather reports. Other savvy computer programs turn these estimates into needs for staff varying by time of day, day of the week, and period of the year. Then the computers fit the available workers into the schedule slots to maximize efficiency, making sure no more workers are paid than are absolutely needed. With this technique, workers' schedules become highly variable and unpredictable. When the number of call centre workers on a given Sunday depends on whether there is a popular football game going on, it is easy to see that George, the most junior call centre worker, isn't always going to work on Sunday. And

George's colleagues don't always know in advance who they are going to see when they come in to work.

An example of what happens when teams are reshuffled comes from Ana María's study of nurses. In the late 1990s, the Québec Ministry of Health put more and more nurses on "on call" status for part or all of their shifts.[10] They were changed from ward to ward and from shift to shift, depending on hospital conditions and expert calculations of how many hours of care each patient would theoretically need. The result: a hospitalized patient who was cared for by "Judith" on Monday morning had less than a 50 per cent chance of seeing Judith again on Tuesday morning. In fact, during the month of our study, half of the shifts on the ward were covered by nurses who worked only one or two shifts on that ward during that month. This heavy turnover had predictable effects on patient care, and therefore on how the nurses felt about their jobs. One complained that being shifted from one ward to another meant that she never knew where to look for equipment or information and didn't even know what was available. Where was the ointment? Was there a film on post-op care that she could show to families on departure? What had families been told about wound dressing? Another on-call nurse said she always felt disorganized and would call the ward after leaving with many details she was afraid she had forgotten to tell them and tasks she hadn't been able to do.

When Ana María observed the assistant head nurses at work, she found that they barely got to speak to patients on their first day on the ward after an absence; they were too busy with paper work and with acquainting themselves with the patients' files. The fact that they were scheduled irregularly and changed departments often meant that a lot of care time was lost. Even a nurse who often worked on the ward spent one third as much time (8 per cent of the time) with patients on her first day back as on her second or third (25–26 per cent of the time). As for an on-call replacement nurse who had not previously worked on the ward, she could only spend 2 per cent of her time on direct patient contact.

Teamwork was nearly impossible under those circumstances because the on-call nurses didn't know each other or the regular nurses, nor did they know the nurses' aides they were paired with. During fourteen days on a typical ward, there were only three shifts during which a nurse was paired with the same nurses' aide a second time.

Right after this study was done, I happened to be in a meeting with the fellow from the Ministry of Health who was in charge of evaluating care. I tried to find out whether patient care was affected by the constant reshuffling of nursing personnel. He was frank: he had no way of knowing. Quality of care was only measured by the number of relapses and a relapse was defined as an incident where a patient was returned to the hospital within one week with the same diagnosis.[11] After the week was over, the patient was treated as a new case. So the patient who went home without seeing the video on postoperative wound care and whose wound got infected wouldn't show up in the statistics as a relapse unless the infection happened in the first few days. And, if I understood the man from the Ministry correctly, even then it wouldn't be counted as a relapse because an infected wound isn't in the same diagnostic category as the broken leg or cancer that brought her in the first time. The costs of breaking up work teams would stay hidden.

Teams are a critical part of most people's jobs, when they are able to form teams. In our experience, most people asked what they like about their jobs rate the interactions with fellow workers as an important part of pleasure at work. Teams are also an important way to learn on the job. Two CINBIOSE researchers, Nicole Vézina and Céline Chatigny, found that it was from their colleagues that slaughterhouse workers learned how and when to sharpen their knives. Having a sharp knife is a requirement not only for doing their job correctly but for preventing injury, since sawing away at a carcass with a dull knife leads quickly to shoulder, arm, and wrist problems.[12] Nicole and Céline were able to systematize what the workers were telling each other and eventually turn their knowledge into a widely used training film.[13]

It is not surprising that the workers liked the experience of having their previously unrecognized expertise turned into a training film. They didn't even charge the employer for their help. I suspect that one reason Ana María is so popular with workers is that she is so quick to recognize their expertise and their ways of supporting one another. Everyone likes being appreciated.

But appreciation isn't just fun. Employers who don't recognize their employees' expertise and work organization pass up important opportunities to make their businesses work better. And the costs of irregular, unpredictable scheduling and breaking up teams are not limited to employers. Employees and their families bear the largest burden, as we will see in the next chapter.

Chapter 6

Home Invasion

When Workers Lose Control over Their Schedules

> . . . periodically, a servant shall be permitte[illegible]
> her friends. . . . At the same time, a mistre[illegible]
> not to bind herself to spare her servant on a [illegible]
> month, as is sometimes demanded. "Once [illegible]
> convenient" is a better understanding.
>
> — Cassell's Household Guide, New [illegible]
> tion, *c.1880s, giving an employer view of atypical scheduling*

A FEW YEARS AGO, DURING A PUBLIC DISCUSSION of store opening hours, a letter was published in a Montreal newspaper. It argued in favour of extended hours, and the writer was affirming his right to buy a veal cutlet at two in the morning. The writer didn't give any special reason why he would need this service. If the owner wanted to open the store at two in the morning and if the customers wanted to be there, why would the government intervene? The government appeared to agree, and stores in Quebec have few limits on opening hours.[1]

Reading the letter, I was brought back to February 1993, the eve of my fiftieth birthday. With Ana María Seifert and Céline Chatigny, I was doing a study of municipal cleaners. Their cleaning of a big sports centre got done at night when no one was around to be disturbed. We needed to observe cleaners during their shift, to understand how they organized their work.

The woman whose work I was observing started at 11 p.m. in the gym. She went up and down and back and forth with an enormous vacuum cleaner while I tried to diagram the pattern she was using. By the hour when the letter writer wanted store personnel to sell him a veal cutlet, I was already weaving on my feet. My son Daood, who can always dream up amazing surprises, showed up unexpectedly at about 3 a.m. with birthday doughnuts, but even they were not sufficient to get me through the night with an intact brain. I finally staggered home at 7 a.m., my notes illegible, with little useful memory of my observations. Daood's second surprise, an 8 a.m. conference call with friends and family on three continents, only produced confusion. I couldn't figure out what was going on or who half the voices were. Luckily, Daood's imagination had extended to taping the call. I later convulsed in laughter listening to the recording of my younger son in India, his girlfriend in Japan, my parents in the U.S. and old friends from Europe, Ontario, and Quebec trying to get through my fog to wish me a happy fiftieth.

I should have known. At sixteen, I had lasted just two weeks in my first summer job as a hospital attendant. As the most junior member of the team, I was on nights, but I could barely function and had to resign. I know that people have different biorhythms and not all find night work as hard as I do. Céline had actually taken reasonable notes on the cleaners' work. As scientists put it, some people are "larks," like me, who wake at dawn, and some are "owls" who can stay up late. But there are fewer owls than larks. When stores and businesses are open evenings and nights, they can't fill their shifts with owls alone. Many of their staff must pitch in, including the larks. The person at the cash register when the letter writer comes through with his veal cutlet in the wee hours has probably not chosen that shift, she is just the employee whose number came up that night. Certainly, the cleaners whose work we observed had no choice; if they wanted a unionized, relatively well-paid job with the city, they had to take the night shift. The cleaners we observed were not much younger than I, and not much happier with the time of day, but there was no question of their asking the young people who used the gym during the day and evening to move over while they vacuumed the exercise mats.

Health effects of shift work are not confined to larks, either. Shift workers in general have a greater risk of injury.[2] Women doing night shift work are more likely to get breast cancer and may also have reproductive problems.[3] And Daood's idea of bringing doughnuts was right on target. I scarfed them up, a food that I, as obsessed with my weight as any middle-aged North American woman, hadn't tasted for at least ten years and haven't eaten since. An injection of fat and sugar was just what I felt I needed to keep me going – possibly why obesity is associated with night shift work.[4]

I don't know whether the letter writer would have cared whether his supermarket cashiers got fat or died young. He might have assumed that they would get extra pay for night hours, compensating them for any lost years or unhappy boyfriends. I don't know, either, whether he had thought about the effects of extended store hours on their family life.

My previous experiences of sleep deprivation had occurred when my children were little. My memories of 1963–68 are foggy and grey. I didn't sleep through an entire night very often, with one or other of my sons waking me shortly after midnight. Like many young parents, I would periodically make a firm decision to tough it out and let the kid cry. This tactic, associated with clenched teeth and a breaking heart, occasionally succeeded in getting my sons to sleep through the night for a few nights. I would then glory in my "victory" and enlighten other mothers with my winning strategy, until a trip or an illness or a change in the weather restored the baby to his original habits and me to my fog.

Recently, watching my sons and their wives go through this in their turn, I remembered how impossible it was to behave at work as if one did not have children. Aside from the lack of sleep, new parents of even the healthiest children are beset with terrifying decisions and lip-biting puzzles. How high a temperature before she stays home in bed? Is it better to have day care near the office (long transport time, difficult bus transfers, weather exposure for the kids) or near home (slower access in emergencies, a longer day away from parents)? Is it wrong to give them a little snack while they're waiting for you to cook dinner if you were late getting home? How can you reach the kids' doctor during office hours when you have the same office hours? How can the school reach you in an emergency if you aren't allowed to access a phone at work? How old do they have to be, really, before you can leave them alone for a half hour?

Can employers expect parents to perform the same as non-parents? If they do, what is the price to the children? To the parents?

My sons and I actually had relatively good conditions during the most taxing years. We are all professionals who, to some extent, controlled our work hours and work days and could make and receive phone calls from work. When we worked evenings and weekends, we were usually at home, where our spouses and children could talk to us if they needed to. But what about those whose work schedules and working conditions are rigidly controlled by others or, even worse, by machines?

I started to learn about how work invades the family life of lower-paid workers in the 1990s, when the FTQ union confederation asked us to study work-family interactions and suggest policy change. Initially, I couldn't see how we could use our major tool, observation, to study this subject. It seemed to me that any family concerns would necessarily be inaccessible during work time since family life was not supposed to be visible at work. But a group of sociologists from CINBIOSE agreed to take major responsibility for the study so I said I would come along and see if I fit in anywhere. We hired Johanne Prévost, another ergonomist, to help out, and we started off by observing work at a call centre. I was struck right away by how horrible the schedules were.

Every Tuesday, call centre workers gave the employer their schedule preferences. On Thursday, the call centre's computer program produced the time sheet for the following work week, considering predicted variations in call volume, workers' choices, and seniority. The workers (if they worked on Thursday) could consult the posted sheet and learn what their schedule would be, starting the following Monday. If they didn't work on Thursday they would have to call in and get their schedule, assuming the supervisor was available to take their call. They could be scheduled at any time between 6 a.m. and midnight. For example, a worker could be scheduled for 6 a.m. to 2 p.m. on Monday, 4 p.m. to midnight on Tuesday, 8 a.m. to 4 p.m. on Wednesday, and so forth. Their days off could be on weekends or not, and their two days off per week could be on subsequent days or separated. Their rest break could come at any time, including forty-five minutes into their shift and seven hours from its end. I later learned that this method is now used to schedule many jobs in the service sector.

We were observing on a Thursday, so we could watch what happened when the workers saw their schedules for the first time. They impressed us with their instant analysis. Seconds after looking at the posting, a worker told us: "I can't work the hours they gave me on Wednesday. I want

Marie's hours but she won't like mine. She does like Jacqueline's. Jacqueline won't like mine either but she'll take Annie's and Annie will take mine. So I'll trade with Annie to get hours I can trade with Jacqueline to be able to trade for Marie's."

However, all changes had to be approved by a supervisor who could only be reached on the phone and, amazingly, the call centre workers had no access to a phone during work hours. There was not even a telephone in the break room. (Mobile phones were not yet in common use, but usually, even now, call centre workers would not be allowed to use them while on the job.) In case of an emergency, a school or babysitter could reach the worker only by leaving a message with a receptionist, who would post it on a bulletin board that the worker could check on her break. Workers told us they were always afraid they would forget to check the board and they would not hear that their child had had an accident or was ill. Because no phone was available at work, arranging shift exchanges had to be done at home: workers spent a lot of their time at home trying to reach the supervisors at their office to exchange shifts (an average of 5.2 attempts per worker over two weeks' observation, with only one successful exchange for five calls).

That Thursday, I learned something else about scheduling and family time that really surprised me. One of the women told me that her friend, the mother of a newborn, now only worked the night shift, midnight to 8 a.m. The idea was that her husband, who worked the day shift, would be able to "babysit" at that time before leaving for work at 8.30 so the couple would save money and the schedule would be more regular. But when did she sleep? I could only imagine her exhaustion.

I now know that this is a common practice. I have seen it recently among cleaners in public transport – the 10 p.m. to 6 a.m. team included three mothers of young children.

Aside from the fatigue, when do couples see each other? Duxbury and Higgins found in their study of 500 of Canada's largest employers that 31 per cent of responding families used a strategy called "off-shifting" to manage their shifts. The two parents work different shifts, with obvious potential consequences for family life.[5] When I asked a union representative about this, he told me that, yes, there were a lot of divorces among couples on shifts.

In order to see how workers actually manage their shifts, my research associate Johanne Prévost thought up the idea of having the call centre workers with young children keep diaries where they would log every action they took to arrange child care during work time. This would be the next best thing to observing them at home. She discovered that they spent a lot of time at home actually working for the company, albeit without pay, in order to arrange their lives around the constantly changing schedule. They were exerting a huge effort to find, schedule, and keep babysitters; the thirty workers we surveyed used a total of up to eight different babysitting resources to fill the slots in a two-week period: spouses, grandmothers, neighbours, day care, other relatives.[6] This effort had to be repeated every week because of the extreme variability of their schedules, which didn't diminish very much even at fifteen years' seniority. Also, from time to time a grandmother would become exhausted or ill, a neighbour would move or get fed up, or a husband's schedule would change, creating a scramble to find other resources. We had no idea how the children who had eight different babysitters in two weeks coped with the constant changes; who kept track of their homework in the evenings?

At the end of the study, we met separately with workers and managers to discuss our results. The managers' language was all about "organization." A woman with children who was "organized" would be able to arrive at work on time, did not need to exchange her shifts often, didn't arrive late. We tried to show the managers some of the effort that went into this "organization," but we were not very successful. Some of the supervisors had had young children themselves, and they had succeeded in organizing their time – why couldn't their employees?

We met with the employees to present our report and that meeting was hard too. We could see why being "well organized" was just not enough. One woman's husband was a hospital worker subject to the same kind of irregular and unpredictable scheduling. The schedules of both changed every week, so the only way to ensure child care was to find someone to fill in when the parents' schedules overlapped. No one but a family member would put up with the consequent variability, so every week the couple would get together with the two grandmothers and try to work out the babysitting schedule. But first one grandmother, then the other, opted out. The grandmothers, who had worked all their lives and were not in perfect health, couldn't live with the constantly changing schedules, and the worker had no idea what she was going to do.

One of the operators committed suicide shortly after our study. We don't know why she committed suicide, but we certainly got a sense that some of the women were desperate. They couldn't hope to find another job that paid as well, certainly not with anything like the same union-controlled pay and vacations. And service jobs everywhere were adopting this kind of scheduling.

We had some suggestions for change, but the call centre workers and/or the employer rejected most of them. The workers were entirely opposed to asking the employer to take their family situation into account; their family situation was none of the employer's business. One woman told us that her friend's availability was conditioned by visiting hours at the prison where her husband was held: did I think she wanted to explain this constraint to her boss? They didn't want more advance notice of the schedule, either. If they were given their schedules a month in advance, that would mean they would have to furnish their schedule constraints and choices for a month in advance, but they were so accustomed to the last-minute system that they couldn't imagine planning that far ahead.

Also, the more senior workers and the union representatives had developed tricks that enabled them to manipulate the computer program. They knew, based on their colleagues' choices, the season, and who was on sick leave, that this week they could get 9 to 5 on Wednesday if they asked for it but that asking for 9 to 5 next Wednesday would only get them the 4 to midnight shift. This expertise would become useless if we were successful in changing the system, so they were reluctant to bet on our improving anything.

We tried again to work on scheduling in 2008 at a unionized, parish-run nursing home, with another FTQ union. Compared to the call centre, schedules were more predictable for senior personnel, but more recent hires (the majority of those with children) were often scheduled on a last-minute basis. I hoped that mobile phones would be making life easier, and they did, to some extent – it was a good thing that the mothers could get emergency messages and check on sick children during break times. However, arranging their schedules still required an enormous amount of invisible, unpaid work. In addition to all the finagling with babysitters and relatives to cover the changing hours of work, there were all the times the system broke down – the car didn't start, the grandmother was busy and, most commonly, the child got sick. Any child in day care exposed to the germs borne by all the other kids gets sick really often, and one of the more agonizing decisions to be made at 6 a.m. is whether or not the child is well enough to go to day care or to the sitter. When the outside temperature is -20°, as happens in Quebec, mothers hesitate before assuming that the growly cough will pass or the tickly throat isn't serious. So they call one of the administrators responsible for scheduling, who is NOT happy to get the call.[7]

Like the situation with waitresses and tipping (chapter 5), the power relations made scheduling more complicated. When we talked to the schedule administrators, we sensed their distress. They needed to keep friendly relationships with the care aides to make their job bearable and also to get co-operation in filling the gaps in their schedules. But they needed to keep from feeling too sympathetic toward workers' family problems or they wouldn't be able to staff their shifts. So they had to be sceptical of workers' stories. They had to decide, several times a day, whether to believe stories of children's illnesses, car breakdowns, broken bones, or relatives' deaths.

It was also the administrators' job to call in replacement workers, in order of seniority and respecting their stated availability. This availability could not be too restricted: a worker was not allowed to refuse to work in a particular department and could only exclude one shift (e.g., evening *or* night shifts) for example. Schedule managers could phone workers at any time, even after the beginning of the shift, and workers could not refuse a shift for which they were theoretically available. And the availability of mobile phones worked both ways. The workers were supposed to furnish the number of their phone and they were supposed to be reachable before and during those shifts, even though they got no pay for availability or for making all their plans provisional. Since people usually have better things to do than waiting for the phone to ring, last-minute calls were not always welcome. It was up to the supervisor whether she accepted the explanation that the employee who missed a call had really suffered from a phone malfunction or whether she had taken evasive action.

The schedule manager explained the system for replacing people who call in sick after hours. "[The night schedule manager] starts with the replacements at about 5 a.m. [When no one answers the cell phone] she leaves a message saying that she has a day shift for her. Then she continues her calls. If in the meantime the lady calls back and the replacement is OK there's no problem and we'll give her the shift."

In other words, the night supervisor is going to wake up several families at 5 a.m. Each of them is going to have to consult with her family to see how to arrange child care, what to do about Tanya's lunch, whether someone can pick up Ronnie from school, then call back the supervisor. But only one of them, the most senior who calls back in time, will get the shift. For her it is worth it, because she gets the hours and the pay. The others have been woken and disorganized for nothing. Is it surprising that they don't like the schedule managers and avoid their calls? In contrast to the other jobs we have examined, the job of schedule manager is a step up. But it is a job right at the lip of the empathy gap.

From the managers' point of view, the women should be able to "organize" their family lives so that the latter become invisible when at work. Once in a while they can be indulged, but not "too" often. The result of this is suspicion on both sides. A health care aide whose sick child had frequent doctor's appointments eventually stopped giving advance notice of the appointments and started calling in sick herself at the last minute. She told us she had done this because of the supervisor's attitude: "You want to be honest, you want to tell them in advance so they won't be stuck. But with my experience, I've noticed that honesty doesn't always pay."

Perfect solutions are not easy to find, since those who work 9 a.m. to 5 p.m. are very happy that services are available outside those hours and not too interested in hearing about the families of those who do the serving. Even those who work the odd hours are happy that their grocery stores and credit card services are open on nights and weekends, and they wish day care centres would open nights. It is hard to think of a way to avoid night work in nursing homes or, in the twenty-first century, to make people do their grocery shopping on weekdays. Children are always going to get sick and cars are always going to break down, so workers are always going to have to be replaced at the last minute.

A clue to some of the ramifications of scheduling came to me from "Lina," the daughter of a friend of mine, who worked as a checkout clerk in a small food market. Lina explained the system to me. Each department (fruits and vegetables, meat, fish, service, etc.) had a manager who was responsible for scheduling under guidelines from the store manager. The managers are responsible for staffing their departments whether or not the workers show up. If there are not enough cashiers or butchers or people who stock the shelves, the customers will be very unhappy. If the managers can't replace an absent worker, they have to work the shift themselves. Since managers, unlike cashiers and clerks, are on weekly (not hourly) pay, they aren't even paid overtime for taking over an absent clerk's or cashier's evening shift. So they are tempted to exercise pressure to get their other employees to replace the mother with the sick child or the father who is taking the car to the garage. A system comes into play that resembles "You scratch my back, I'll scratch yours." If the manager calls you to work tonight and you don't accept the extra shift, don't expect him to let you go home early next Saturday evening to watch your daughter's hockey game. If you refuse too many extra shifts (or evening shifts, or shifts with the colleague who has body odour), watch out! You have a good chance of finding yourself scheduled to work every closing time for a week and get out after the buses finish running.

Lina, a pleasant young woman with a cute sense of humour, appeared to spend a good bit of her fantasy life thinking of ways to get revenge on her manager. She eventually decided to go to beauty school and become a pedicurist. Her small grocery store was not unionized and employees like Lina had little recourse against unjust scheduling. But even large unions have had trouble figuring out how to make schedules fair. We were asked by the FTQ union women's service to look at scheduling in a large retail chain with many franchised stores. When we went to see the human resources people at the head office of "Qualiprix,"[8] they were delighted to see us. They were having a lot of trouble with high turnover: 80 per cent of their workers left every year. They were hoping that scientists could help figure out how to make the chain's employees happier, and they were impressed with our idea that maybe their scheduling software could be improved so as to take workers' family needs into account. They were very pleased to talk about their software with the university-based consulting firm we brought with us.

Our interests didn't dovetail perfectly, though. We noticed that "M. Lejeune," the human resources guy, kept referring to specialized workers and assistant managers, while we were thinking about cashiers and stock clerks. Their idea of workers' personal emergencies was students faced with an unexpected examination, whereas we were thinking about toddlers with chicken pox. Nevertheless, they collaborated generously with the study we proposed.

We observed the work in two stores and asked workers in seven others to answer a questionnaire. Administering the questionnaire was quite an experience. We visited the individual stores and arranged with the local owners to explain the questionnaire to a few workers at a time; we stayed for three days in each store, long enough to reach most of those regularly scheduled. The stores varied widely: one had a big, fancy conference room where employees could fill out the questionnaire on comfortable, upholstered chairs. Others were small or just badly set up, with us and the employees filling out the questionnaire all squished in a tiny, dirty staff room getting in the way of those trying to take their breaks or get to the bathroom.

In all the stores, the employees were enthusiastic about the questionnaire. Almost all those asked to respond said yes, and many thanked us warmly for our interest in their schedules. I became afraid that we were creating high expectations that schedules would change, while the most we could hope for in the short term was to change owners' attitudes.

Hanging out in the stores, we got a fair idea of what those attitudes were. In one store, I shared the owner's large office with a table in the corner where the workers could fill out the questionnaire. Although I was glad that we didn't have to invade the employees' break room, I was less pleased with the arrangement when he routinely greeted the workers with "Here's Bill (or Anne, or Rafael) – he's one of our grant-supported workers." This was meant as a funny joke, since the grants came from government support to hire the mentally disabled. The employees, presumably gritting their teeth internally, smiled and nodded at the joke, every time.

The owner offered repeatedly to fill out the questionnaire, since he said his work hours were unlimited. He did work Saturday. However, I didn't see him leave after 6 p.m., and, unlike his employees, he took long lunches outside the store. But I do believe that he sincerely thought his schedule was as taxing as that of the employees, because control over scheduling is only visible when you don't have it. And the degree to which he could freely express his contemptuous attitude toward his employees showed me that he felt pretty much in control.

This owner was the only one whose store was not unionized, so possibly he felt freer to treat his workers cavalierly. But attitudes of other owners were not too much better; most seemed indifferent to the workers' needs, and one said he would only consider family responsibilities if the worker was legally married (the situation of only two-thirds of all cohabiting couples in Quebec). To be fair, managers were worried that if family responsibilities had to be considered during scheduling, they would be unable to ensure that workers with the required technical skills were in the store at the right times.

The answers to the questionnaire depressed us. These employees get their schedules on Thursday or even Friday, for the week that starts the following Sunday. This means they don't know until the end of the week whether they can go away for the weekend, whether they can use the concert tickets they have been offered, whether they can ask that attractive brunette to their friend's party on Sunday night. And in fact, more than 80 per cent of them can't do much because they work at least one weekend day; half had worked both days of the previous weekend.

Sylvie is a middle-aged cashier who is the sole financial and logistical support for her two aged parents. She arranges their medical appointments, their shopping, and their banking and accompanies them on trips to the dentist or the hairdresser. After fifteen years' working at the store, she never has the full weekend off, and she never knows anything about her hours before Thursday. Here is her response to a question about desired hours, a request for the working conditions that most middle-class people enjoy well before fifteen years' service: "I would like to have always the same days [off] and that they would be two together. Have the same days and the most possible hours per day. Have my weekends [off] the most often possible. Know my schedule at least one week in advance. . . . [Please try to] improve the conditions of employees with low salaries who have to accumulate so many [paid] hours to take care of their families' needs, because they have no free time with certain schedules."

In other words, Sylvie is faced with a paradox. Like other workers with a family to support, she has to work lots of hours because her pay is so low. But the work schedules are so unpredictable and demanding that, given her family situation, she can't work as many hours as she needs.

Sylvie wrote us long notes all over her questionnaire, and she was not the only one. Some of the notes we saw were short and to the point: "Grrr [design of a scowling face] I hate working Fridays!" Some of them were resentful: "The employer often doesn't like our opinions or ignores them when hours are handed out."

Qualiprix Stores was actually a relatively good employer. The jobs were unionized and the employer was quite concerned with retaining its workers. Its human resources department came up with new ideas to make the workers happier and even put some into practice. But many people with family responsibilities couldn't afford to take a job at Qualiprix Stores. The schedules were impossible for them.

When we analyzed our results we were surprised that only one-third of workers reported finding it hard to reconcile work schedules and personal needs. Then we figured it out. Only 17 per cent of workers had young children or dependent relatives, compared to about 40 per cent of Quebec workers; 71 per cent had no family responsibilities whatsoever and lived alone. In other words, if you had a family, you would go elsewhere to look for work. And those who said they had trouble reconciling work schedules with their personal lives were among the same 56 per cent of employees who admitted dreaming about finding another job.

This morning's e-mail brought me a call for papers for a scientific journal in psychology. It said in part: "[Our proposed journal issue aims] to help employers as well as employees to better use the skills and resources (e.g., time management, communication, sense of organization) developed by work-family balancing. In fact, is it possible that our personal and professional spheres of life could coexist in a . . . necessary compromise, or, on the other hand, constitute sources of mutual enrichment?"

Reading the retail employees' answers to our questionnaire on work hours, I heard little about enrichment but a lot about compromise. In their comments, workers offered to accept bad weekday shifts in order to get two weekend days from time to time; they offered to work late if their shifts could be regular; they would work long hours on some days if they could have their weekly days off two days in a row. Although some researchers have indeed found that women's mental health is improved if they are active in both work and home spheres, very little of this scientific literature on stimulating work and enriched family lives comes from service workers at the bottom of the pay scale; most concerns professionals and managers.[9] Also, almost none of it has described the actual workplace conditions that make work-family balancing nearly impossible, like those in the retail stores we studied.

Researchers who have not studied such workers close up seem sometimes to have no idea what they are up against. My engineer colleague had brilliant ideas on how to introduce family needs into Qualiprix software, but his vision of family needs came from his own family: two young, well-off professionals with healthy children and helpful, retired, well-off grandparents. When I read him Sylvie's cry for help with her problems caring for aging parents, his response was, "She really should get another job." I'm not sure what other jobs he thought were available in the suburbs for an uneducated, middle-aged cashier, but I'm confident that Sylvie had already gone through that search again and again.

We presented our study to the employer and, for the first time in my career, high-up executives were interested in my research results. They were quite impressed by the fact that workers who found the schedules hard were the ones who were thinking of quitting. Unfortunately, like M. Lejeune, the executives talked a lot about the skilled workers, and not at all about the cashiers. They worried about the fact that students were no longer willing to work on weekends but they appeared to sleep through the data on the effects of irregular schedules on young children. In fact, the union representative advised me to leave that part out next time I made the presentation – she said the managers were looking too bored.

We had brought our legal experts with us and together we suggested that if the stores wanted to make the employees happier without losing out to the competition, it would be a good idea to lobby for government regulation so that opening hours would be limited for the whole retail sector. This suggestion also elicited a dead silence.

I learned from this experience that empathy is a luxury for those who have to keep a store running and profits coming in. M. Lejeune and even the consultant weren't trying to put themselves in the place of the call centre operators' children or in Sylvie's place – there was every reason for them not to do so.

Is it any wonder that some questionnaire respondents were sceptical about our study? One young man described the empathy gap very well: "I just don't know what use it is to fill in this form since I know very well that the Université du Québec doesn't give a shit that I think I don't have enough [paid] hours in my stockroom at Qualiprix Stores Saint Jeremy Québec."[10]

I'm not sure what use it was for him to fill in the form, either. His boss was the guy who consistently disrespected his employees and I am sure that he never was allowed access to our report or to the study results. Sometimes the empathy gap is just too wide – it is unlikely that UQAM or any other university will succeed in changing his work schedule.

After a flurry of interest that succeeded in changing the work schedules in one store, M. Lejeune called me. Sorry, he said, but we have just bought a new chain of retail outlets. The people at human resources have no time to go on with your study.

During the week before Father's Day this year, eight journalists called me for interviews on work-family balancing for fathers. They had heard from my funder that I knew something about the work/family interface. But they all wanted to know whether fathers were stressed, their attitudes toward their families, things I know nothing about. Some of the women wanted to know whether fathers were doing their part at home, something else I know nothing about. I had a hard time getting them to listen to me about elements of the work situation that make it hard for fathers (and mothers) to be with their families. Finally, one young man caught on. Oh, you're saying it's the system that's wrong for fathers. Yes, I said. The system.

Chapter 7

Teachers and Numbers

They're cutting my wings off!

— *Secondary school teacher in Montreal*

FOR A LONG TIME, THERE WAS a billboard in downtown Montreal that I passed every day. The huge letters said "What gets measured gets improved." People reading the sign were invited to contact a consulting firm to improve their business performance. As the *Business Dictionary* puts it: "What gets measured gets improved. What doesn't get measured doesn't get improved. Measure everything important."[1] At that time, I was engaged in a research project with a master's degree student, Jessica Riel, on the working conditions of secondary school teachers.[2] As our analysis progressed, I had ever-stronger urges to throw rotten tomatoes at the billboard.

Some things are just not measurable. I have a joke with the love of my life, who also has a background in the natural sciences. When he tells me he loves me, I ask how many millicupids. Then we argue about who loves whom more millicupids, and we each try to prove that we are producing more millicupids. We have devised many methods, not all reproducible in public, for measuring the number of millicupids. But this is our way of being funny, see? Unfortunately, sometimes people who read the *Business Dictionary* don't get the joke, and try to measure unmeasurable things, with horrible consequences.

In Quebec right now, a man named François Legault is doing well. He has formed a political party and promises to solve a lot of our problems by using the approaches he developed while heading an airline. For instance, he wants to improve the educational system, because not enough students finish high school before they are twenty years old. His solution is that salaries for teachers should be raised by 20 per cent, and, in return, teachers' performance should be evaluated twice a year by province-wide examinations of their students and polls to determine parents' satisfaction with how their children are doing. Teachers should no longer have permanent contracts, so that those who perform badly on Legault's measurements can be fired.[3]

There are obvious problems with measuring by these methods. How will M. Legault compensate for the differences by school and class in clientèle, in parental literacy, in resources given to schools and teachers? How will he deal mathematically with the number of disabled students or slow learners per class, the number of students whose first language is not French or English, the number of parents who have no idea what is happening at their children's school, who are angry with the school for reasons having nothing to do with the teacher or who have no qualifications to evaluate teaching? But the *Business Dictionary* is right about one thing. It is very likely that what is measured will improve. If their jobs depend on it, probably quite a few teachers and principals will arrange things so these indicators will show progress. Will they discourage poor performers from entering their school? Will they ignore material that isn't on the provincial examinations? Will they have parties for their students' parents? Will some of them even give the students little hints on the examination questions?

It is understandable that parents and, therefore, the government are concerned by the educational system. There is nothing more important to most parents than their children's future, and many feel that their child is not sufficiently helped, appreciated, or understood by their teachers. The idea of doing away with job security for incompetent teachers probably gets a lot of votes for M. Legault, but our studies of teachers' working conditions have shown us that quantifying their performance is not easy and may lead to disaster.

How to improve teachers' working conditions while keeping high-quality education? Well, we could try the Finnish approach – lower class size dramatically and make teaching one of the most high-prestige, attractive professions.[4] However, M. Legault wants to lower taxes without spending more on schools, so that solution is out.

Using salary as a metric for employer satisfaction and therefore job performance has a few flaws, too. A study has shown that higher pay for teachers is not associated with better performance by students.[5] More importantly, I have never met a teacher for whom salary was the major source of gratification they got from their job. I don't deny that teachers are paid better than retail employees and cleaners, and I'm sure they're happy about that. But, as elsewhere in the public service, salary is not the major currency in which teacher prestige is expressed.

Let's get back to the millicupids for a minute. When we studied primary school teachers, we found that they were motivated by love. "It's the most beautiful profession." "We feel appreciated by twenty-seven people!" they said. In fact, a better way to say it is that love is a necessary tool of their trade; they couldn't teach if they didn't love their pupils. They explained to us that their jobs consisted of taking "a journey" with twenty-five to thirty new children from September to June every year. In September, they bitterly missed the children from the year before, and the new ones were strangers. As the year went on, they came to know where each of the children was on the path to education, their needs, troubles, and abilities. They slowly grew able to move each one of them along their path. And they often described their behaviour almost as seduction; they tried all sorts of strategies to motivate the children. "I got him!" said a fourth-grade teacher who had succeeded in calming and reassuring a troubled nine-year-old long enough for him to finally understand multiplication.

I came to understand how the teachers felt about their students when one explained to me how she had worked for months to change the attitude of an unhelpful parent: "I did it for the child . . . and she has two younger brothers coming up so all of them will profit." This teacher told me she had trouble explaining to her husband, who felt she was overinvolved, that it was a good idea for her to "get some emotional distance" from her pupils. But, she insisted, she wouldn't be able to do her job without love.

Can we measure the ideal emotional distance in millicupids? When we observed elementary school teachers in eight classrooms for a total of forty-three hours, what could we count? We did end up measuring several indicator variables: the length of teachers' sentences, the duration of their eye movements, the length of time their backs were bent low over the pupils' desks, the number of phone calls and other contacts they made outside of class time in order to help distressed or abused children, the temperature, humidity, and noise level of the classroom.[6] For example, we were able to find out that the average length of a teacher's sentence in grade 1 was 8.7 seconds and in grade 4 it was 14.4 seconds. It seemed to us to be a good way to show how the attention span of the children got longer, and emphasized the many ways the teachers adjusted their explanations to the state of the children in the classroom.

So measuring did serve to make visible some aspects of teachers' skills. But could M. Legault use those numbers? Could he turn sentence length into a component of the equation determining teachers' pay? Would he want them to be longer or shorter or more variable? Did they really reflect competence? We might be able to do something with measurements of temperature and humidity, which were outside desired levels in all classrooms at least some of the time. Conceivably, the length of time outside the comfort limits could be used to rank schools in some way, but that wouldn't be useful for evaluating teachers. Maybe teachers could get points for keeping a pitcher of water available – but then some didn't have pitchers, saying the pupils were always getting up to get water, which upset the concentration of the others. Which was right? Who should get the raise?

Using numbers to control aspects of teaching has a chequered history. For example, counting hours per subject has been used for some time to allocate class time in elementary schools, but we weren't convinced it was the best idea. We had been quite impressed by how the teachers could follow the progress of twenty-seven children at once, keeping a mental register of each one's small victories and defeats at the hands of grammar and mathematics. But these otherwise competent adults kept getting confused by what day it was. The suburban school where I observed was on a six-day week: the first school day in the year was Day 1, the next was Day 2, etc., so that successive Mondays were (for example) Day 1, Day 6, Day 5, etc. As far as I could find out, the reason for this was that a certain number of minutes per year were assigned to each subject area, and the number of minutes didn't fit neatly into a five-day week. Also, holidays have a nasty habit of falling on Mondays and Fridays, so setting up a five-day regular schedule would mean that material scheduled for every Monday might miss some minutes. Rather than letting the teachers work out how they would cover their material, the school board scheduled the time slots on a six-day basis, with every minute scheduled and counted. Of course, children being children, the minutes that a teacher could actually spend on the prescribed subject matter during each scheduled time period varied widely. One day, a child mentioned during a lesson on gerunds that his grandfather had died and the teacher (as good teachers do) interrupted the lesson to talk about it before folding the gerunds into the discussion. Another day, a little girl burst into tears during the arithmetic period and the teacher (as good teachers do) took time to find out why and comfort her. Kids throw up in class, ask unexpected questions, catch on faster than expected, are distracted by passing flights of birds, etc., etc. And we watched several good teachers profit from these events to teach something. Scheduling minute by minute in elementary school is ridiculously inappropriate and ineffective as far as determining how teaching time is apportioned.

But scheduling did have some effects on teachers' work. Many mornings when I walked into the staff room, there was a discussion about what number day it was. Sometimes a teacher had prepared the "wrong" lesson or mis-scheduled an appointment for the wrong free period. I myself got mixed up about when and whom I was supposed to observe. And the six-day schedule wrought havoc with visits from learning specialists and other outsiders, whose work week was scheduled Monday to Friday.

When we studied secondary school teachers, the "minutes" approach had just been extended. Although secondary schools had long been on six-day or even eleven-day scheduling, the teachers' preparation time had still been under their control. But in 2005, the government decided to take a more directive approach to teachers' workload. It was known that teachers spent time outside class correcting student work and preparing lessons; we had found that primary school teachers logged an average of 16.1h/week of such activities in addition to the paid work week.[7] They cautioned us against thinking that those activities defined their work week. "There are three schedules: the one you're paid for, which is twenty-seven hours. The one you do, planning, correcting, and make-up, which might be sixteen more hours. And the constant thinking about it, which is all the time."

I don't know how the figure was arrived at, but the 2005–2010 collective agreement (Chapter 8 paragraph 2.01) introduced a requirement for high school teachers to spend five supervised hours per week at school doing "personal" work, defined as preparing classes, correction, reporting on student progress, working with other teachers or school professionals, and dealing with other school-related tasks. In the school commission about which Jessica Riel, an ergonomist in training, did her M.Sc. thesis, they had to fill out "minute sheets" (time sheets) in advance, where they indicated the task they would do, the precise time of day when they would do it, and the place at school where it would be done under conditions where their physical presence could be verified. If necessary, with due notice, the work could be done elsewhere, but any change of place had to be approved. Instead of doing their work quietly at their homes, which most had furnished (at their own expense) with computers, adequate desks, telephones, and a quiet environment, they usually had to fulfil the "minutes" requirement in the staff room, which was crowded, noisy, and had only three computers for the entire staff. Phoning parents or social workers could only be done officially by carrying the single phone outside the staff room, assuming the teacher could get hold of it. And of course you couldn't rely on that phone if you needed to be reached.

A French teacher told us, "You want to do the corrections, but there is always someone who asks you a question or teachers who are talking. Sometimes the phone rings, you'll answer it because you're nearest. It's hard to concentrate." Most of the five hours' work at school was not deducted from the sixteen hours at home, it was added on. So the teachers' unpaid work week was lengthened because of the government's desire to measure and control the work they were doing voluntarily. But, for the teachers we talked to, the worst part of the "minutes" system wasn't the overwork, it was the insult. An art teacher said, "All this monitoring . . . We feel like we're being treated like schoolchildren." A French teacher put it more poetically: "They're cutting my wings off! I give myself body and soul and they aren't satisfied."

The school principal, on the other hand, liked the system very well. "It gives better performance. It's about recognizing the teacher's assigned tasks. I'm happy that there's that recognition in the collective agreement: 'You do more but we recognize it.' They [teachers] can plan their time and it allows them to get well organized." In fact, many studies have shown that teachers in Quebec and elsewhere are stressed and overworked.[8]

In the preceding chapters I was talking about our experiences with workers whose prestige is low due to their lack of education and their perceived low level of skill. We would have expected that teachers would be treated with more respect, since they are well educated and trusted with our children. Decision-makers and middle-class parents should empathize with teachers more readily than with cleaners or checkout clerks. Strangely, teachers are often the objects of the most aggressively contemptuous treatment.

Irène Demczuk, a sociologist, did a study of newspaper coverage of teachers while working at CINBIOSE.[9] She searched four Quebec newspapers (*La Presse, Le Devoir, Le Soleil* and *La Tribune*) for two months for articles about public school teaching or teachers. She found 149 articles and concluded that this meant the public was really interested in teaching, which didn't surprise us. She identified six themes in the articles: working conditions, teacher training, program reform, pedagogical days, religion in schools, and accomplishments of teachers. In two categories, the various classes of opinions (editorials and letters) were more frequent than news content: pedagogical days (days when school is closed but teachers go to work for training or planning sessions) and teacher training. Pedagogical days were the focus of an especially hot debate during this time, because the education ministry wanted to reduce the number of days. During the debate, teachers were often represented as working little ("only 200 days per year" and "only 27 hours a week"). Days used for teacher's meetings and professional advancement were called "leave," implying wrongly that the teachers do not work those days. The days working outside the classroom were sometimes described in terms that made them sound like personal growth seminars rather than professional activities. Teachers were represented as wanting to avoid teaching. "Don't Feel Like Working" was the title of one feature article describing a struggle to keep some days for professional training.[10] "A group of teachers manning the barricades to keep from spending three more days with their pupils" was a phrase from another.[11] While it is understandable that working parents resent having to make arrangements for child care during professional days, it is hard to know why that energy is turned against the teachers rather than towards ensuring substitute child care by the school commissions.

The teachers come up against the same negative attitudes when they meet their pupils' parents. One first-grade teacher had to explain to a sceptical father that skills were needed for her work, it wasn't just babysitting. He only believed her after she let him teach her class for an hour. He lasted ten minutes before admitting he had no idea how to control a class of six-year-olds.

We got a dose of these attitudes ourselves at the annual CINBIOSE research conference when Jessica and I presented results of a study of secondary school teachers to other CINBIOSE researchers. Our study was about their terrible working conditions and we thought it would inspire sympathy for them.[12] As with our research on primary school teachers, we had found that classrooms were noisy, dirty, and badly heated. As before, we noted that teachers were on their feet all day and told us they had aching backs. In addition, we found that, in some high schools, teachers didn't have enough support from the school principal when they tried to discipline students. But the discussion of our research results got sidetracked. Instead of considering the teachers' working conditions, the audience, all public health researchers, shared stories about the errors, stupidity, and incompetence of their children's teachers. All the other presentations were discussed in scientific terms, but not this one.

Stranger still, the teachers seem to share the empathy gap with respect to their own profession. Our presentations to the teachers' unions are the only ones where we have major difficulty keeping the workers on topic. For example, we were asked to make a presentation to the union on precarious employment among adult education teachers, 86 per cent of whom were hired on a short-term contract or casual basis, even after fifteen or twenty years on the job. We had a lot of topics to cover: their struggles to survive without economic security; how they coped with working in three or more schools at the same time, with no assigned office space; legal disputes over contract workers' rights to the materials they developed. We had seen major government policy changes that made the demand for teachers bounce up and down, budget changes that arrived after the school year had begun and unforeseen decisions that suddenly lowered the age of entering students from adults to adolescents. And we had seen the effects on teachers, who waited by the telephone anxiously at the beginning of each term, who still couldn't buy a house at the age of forty-five, but who worked extra hours to help immigrants and transgressed the rules to protect poor students.[13]

We never really got to tell the union meeting about any of this, because most of the time scheduled for our presentation was taken up with a discussion of school buses at a primary school. We listened, puzzled, as the adult education teachers united with the others to discuss how their union could get more buses and change the bus schedules. We couldn't understand how the teachers' work was affected – and in fact it wasn't, at least not directly. The teachers wanted the union to intervene on a question of children's safety and comfort, and they wouldn't let go until the union executive agreed, even though their demands had nothing to do with unions' traditional purview: improving workers' lives.

Dorothy Wigmore, a union educator, tells me she had the same difficulty getting teachers' union representatives in western Canada to think about the violence they experienced from parents and students. In a union health and safety training meeting about bullying at school, teachers spent their time talking about resources to prevent students from bullying one another and couldn't get interested in their own experiences of being bullied.

It's not that the teachers are indifferent to their own welfare. But their welfare is closely tied to how the children are feeling and their ability to teach is conditioned by their love for the students. Without love, they will lose their patience and their capacity to teach. Moreover, if the air in the classroom is too dry, too warm, or too cold, the children are uncomfortable and restless. If they are being bullied, the classroom atmosphere is tense. If the students feel that the teachers love, protect and appreciate them, they learn more readily.

More broadly, teachers get their job satisfaction and self-respect from giving to their students. In the younger grades, they love and care for the children and they want them to learn. As the students age, their relationship with teachers has a cooler temperature, but the teachers still passionately want the students to progress.

So the teachers' propensity to concentrate on the well-being of students is understandable. But it might be better – for them and for the pupils – if the parents, the school authorities, and the government backed them up. Instead, the government approach appears to resemble bullying.

During a time I was studying teachers, Pierre brought a colleague to dinner whose wife was a school principal. I was delighted with the opportunity to get her perspective on the mental health crisis among teachers. More than one teacher in five was quitting the profession during their first five years of teaching, and many were retiring early, citing burnout.[14] I asked the principal whether this was her perception of the situation. "Yes," she said, "they're dropping like flies. We have [workers' compensation] cases all the time now." So, I asked, "Do you have a program for prevention?" "Of course," she replied. I was thrilled and asked, "What's your approach?" "We fight each case," she replied proudly, and proceeded to explain how her school board had hired lawyers to make sure the teachers' stress would not be recognized as a work-related illness. Many jurisdictions have gone further, and specifically exclude teachers' stress, by law, from workers' compensation. I wonder why they would think that would be necessary – could it be that current conditions in the profession are stressing a lot of teachers and potential costs of compensation might be high?

The contradiction between her observation that teachers were "dropping like flies" and the lawyers trying to establish that teacher stress was unrelated to work seemed to have escaped the principal. She reminded me of a scientist I had met a few years earlier who had invented a system to calculate how much of a particular worker's cancer could be attributed to his work. When I ran into "Professor Numbers" at a seminar, he told me employers were calling him in to testify at workers' compensation hearings. They found him useful, because his system allowed them to contest the requests for compensation, saying that the cancers weren't caused by the workplace. He mentioned that one employer was keeping him very busy, since a number of workers at that factory had bladder cancer. He complained that he was spending so much time testifying that their cancers weren't related to the workplace exposures that he was neglecting his own research. But, he said, he had been quite successful in winning the cases. I found it a bit hard to understand: if the cases of this rare cancer weren't related to the workplace, why were there so many cases at that factory?

It is numbers like these that make me mistrust the "Business Report" approach to work organization. Whether calculations of how many hours of nursing care per day are needed to keep patients from getting sicker or how many minutes of arithmetic lessons will suffice to teach long division, the numbers seem to take on a life of their own. Instead of being a rough guide to staffing, they tend to become a standard for performance. And they rarely include any wiggle room, to allow for the extra time it takes to put on support stockings when the patient is obese or in pain, the lonely, garrulous patient, or the class where half the children don't speak the language well.[15]

Numbers that describe work can be helpful to illustrate a problem, such as when Ana María counted the number of times the bank tellers were interrupted. The numbers showed that the system for sharing information on bank procedures was faulty and should be improved. Numbers can even be helpful to follow progress in solving a problem, as when we counted the proportion of physically demanding tasks that were shared in the nursing homes. But the numbers are only useful when the person using them thoroughly understands the work process and can be sure that the numbers reflect what they are supposed to.[16]

When I teach young ergonomists, one of the hardest points to get across is that using numbers doesn't make a study more objective. Using numbers makes a study no more or less objective than using words. It is hard for students (and many professors) to understand this because employers, unions, and governments insist on numbers.

For example, one of our graduate students proposed a study on pain among breast cancer patients. He proposed to use an algometer to measure the pain. An algometer is a small apparatus that measures pressure. The experimenter applies a growing amount of pressure to the subject's body until the subject pushes a button to signify the level at which she starts to feel pain. The pressure at that point is recorded as the pressure-pain threshold (PPT). So the hand or chest of a patient who feels pain at 135 kilopascals (kPa) is more sensitive than one who doesn't push her signal until 400 kPa.

These numbers are useful, because they can be followed over time, compared between genders, and used to evaluate drug effects. Numbers are easier to compare and analyze than words. We have used the PPT on the bottom of workers' feet ourselves to see how much sitting per day is helpful to prevent foot pain.[17] But where I got into an argument with the student was over whether they were objective. Why would it be more objective for the patient to push a button than to say "Ouch"? The patient who says she is feeling better or worse is the same patient who chooses when to push the button. And in fact the choice of PPT rather than just asking the patient how her pain is evolving is at least as political as scientific. It is a lot easier to publish a paper in a scientific journal, to get a grant, to convince physicians, to persuade drug companies with PPT figures than patient reports. But it is no more objective. In both cases, the results depend on how the patient is feeling at the time, who the experimenter is, what time of day it is, etc.

By analogy, watching teachers in a classroom for an hour gives a certain kind of information, asking the principal to evaluate the teacher gives another, and testing the students gives yet another. The results of observation depend on the time of day, the attitude of the observer or the school principal, the subject being taught, air quality in the classroom, the number of problem children, and a myriad of other things. The test results depend on the choice of questions, the time allotted, the air quality in the testing environment, the teachers' working conditions, and many other things. Using a single number such as a test score gives the illusion that the evaluation is objective and related to the desired outcome, but it is just an illusion.

Unfortunately, teachers have few ways to make the nature and constraints of their work known to their school heads, the public, and the government. Like other workers caught up in the day-to-day needs of their jobs, they are not always conscious of why they feel overwhelmed or inadequate, of where their difficulties stem from.

Being an outside observer has enabled me to appreciate the challenges involved in these jobs, and being a professor has given me the credibility to make them a bit more visible. But not too many scientists have had this kind of opportunity to develop synergy between their knowledge and experience and that of workers. In fact, I think it takes a special set of circumstances to pull the daughter of a vice president of General Instrument Corporation across the empathy gap. So I need to explain a bit about the years of my scientific training and how I found myself commuting back and forth across the gap.

Chapter 8

Becoming a Scientist

MICHELINE CYR, ONE OF THE GENETICS STUDENTS who helped me study the factory workers exposed to radiation, is the daughter of a less than successful window washer and a housewife who cleaned people's houses when her arthritis allowed her to get around. When I first met "Mimi," she was unprepossessing – softly spoken and inexpensively dressed. She later explained to me how she felt ill at ease at the university and had trouble understanding the fancy language of the professors and even her fellow students.

Mimi and her husband, parents of a young boy, had worked in dime stores and factories for several years before accumulating enough money to start their college education. Despite working and caring for her child, Mimi accumulated a perfect 4.0 average as a biology student. She went on to rank first in Canada on the competition for scholarships in cancer research, and quickly finished a master's degree in molecular biology under the direction of the head of the Montréal Cancer Institute.

She tells a funny story about how her life course then turned a sharp corner. She was starting her Ph.D. program in public health when her family had a Christmas gathering. One of her aunts asked her how she was doing, and Mimi explained that she was still in school, getting a doctorate. The aunt tried to hide her amazement and consternation that Mimi was still in school at the advanced age of twenty-five and said tactfully, "Don't worry, honey. Lots of people have to repeat their years but they finish school in the end." The incident made Mimi reflect on the distance that separated her current life in academia from her family and childhood friends. Some months later, she quit the doctoral program and became the night counsellor at a refuge for street women. She now heads the refuge and has helped to shape the way these services are provided all over Quebec. Her originality and drive have improved the lives of many women.

I try to keep in touch with all my favourite students, so I follow Micheline's career. To me, she is a heroine. But I can only admire her and not emulate her. I have come to realize I could never have worked in the shadows, never done the daily slogging away at household tasks in the refuge, never handled the frustration of seeing women go back and forth between addiction and recovery, never have accepted all the adaptations imposed by her difficult clientele. Listening to her incisive comments about government services, I realize how her background, including her childhood friends and her aunts, has made her able to assess the impact of government programs on people authorities never see or hear from. In her daily professional practice, she is helping bridge the empathy gap between government officials and those they are meant to serve. I wonder whether it is possible for scientists to make a similar contribution to improve the lives of low-paid workers without leaving their own jobs. Could Mimi have stayed in science and still changed life on the street in Montreal?

My own journey is different from Mimi's. I was the child of a vice-president of a multinational corporation and a left-wing artist. This chapter will describe how I went to elite universities, becoming a Harvard student, a McGill University Ph.D., and eventually a university professor and director of a well-funded research centre. My life has been a lot easier than Mimi's, right from the beginning. But even before meeting Mimi in 1977, I was always asking myself what use was being made of my science. Because of my mother's politics, I was aware that my expensive education came from the work of the women who put the coloured wires into the radio sets, and I am still asking myself if they have gotten anything back.

In high school in Massachusetts in the 1950s, I knew girls didn't become scientists. My counsellor explained to me that the fact that I tested high in math was a kind of contamination from my being good in English – it didn't mean I was *really* good at math. When I got straight A's in physics, I didn't expect to be selected for the special physics camp and I wasn't surprised when the boys went off without me. And no one thought it was important for me to take chemistry and biology in high school, so I didn't.

When I got to Harvard University, even those of us majoring in social sciences were forced to take two terms of some kind of natural sciences.

Like the other girls, I gritted my teeth and took a history of science course. The course was filled with other social science students who didn't want to be there, which inspired the guy in charge to spend every class reading aloud to us from his forthcoming book. It was so boring that the usually sedate Harvard men threw paper airplanes during class, knowing that the professor would never look up from his book and notice.

My second natural science course was different. The only course available during my final term was a scary-sounding biochemistry course for non-scientists, taught by Leonard K. Nash. He was an amazing teacher. It was 1962 and research results were coming out almost daily on DNA, the chemical that had just been found in genes. Professor Nash brought us virtually to the lab bench and made us understand the process of discovery as it was happening. It was really exciting to learn how twisty little molecules could carry the information for complex cell behaviours.

But, fascinated as I was, I never dreamed that the actions of the smart men I was hearing about had anything to do with my career. It took a feminist to help me make that link. Right after graduation, I gave birth to my son Daood who immediately put his mind to doubling and then tripling his birth weight. I spent almost all of my time nursing him, with a book propped up on my rocking chair. In the fall of 1963 that book was *The Feminine Mystique*, which had just come out. Betty Friedan's book gave me a new idea, that women could be scientists, and I immediately signed up for night courses in physics, chemistry, and biology. I liked them, did much better than I expected, and prepared for a career in science. I thought that medical school might be appropriately situated between science and my bachelor's degree in social science. Naïvely, I wrote to several medical schools in 1964, asking them if they accepted women with children. They all replied that they did not.

I went back to see Professor Nash, who spent time talking with me and encouraged me to apply to graduate school. I wanted to be a chemist just like him. My then-husband, a Muslim, applied to study at McGill University's Islamic Institute, and we ended up in Montreal after a whirlwind visit charmed us with the city and the Institute. My mother-in-law came from India to babysit while I took as many classes as I wanted in preparation for graduate school in chemistry. Unfortunately, McGill's chemistry department was not its strong point and Professor Nash had spoiled me. The teachers were old-fashioned and unexciting and it took no time at all for me to get into trouble with them. After I had denounced them in the campus newspaper for not showing up for class as often as they were supposed to, they made me aware that I would not be welcomed back for a second year.

I had made one friend among the chemistry students, Danielle Saint-Aubin, who did me two favours that pushed my career onto its final track. She was perfectly bilingual and introduced me to the "other" (French) culture in Montreal. After a year immersed in English-language McGill, I was amazed to learn that 83 per cent of the population of Quebec spoke French as their first language. I started to speak French with Danielle, setting out on the road that led me, ten years later, to a job as professor in the biology department at the Université du Québec à Montréal (UQAM).

One day in the fall of 1966, Danielle did a second good deed. She brought me to listen to her favourite teacher, John Southin, a young and dashing genetics professor. Like Professor Nash, he was wonderful at explaining how genetics actually worked. He talked not in formulae, like my chemistry profs, but about mechanisms, *how* the DNA worked to govern cell behaviour. He made us understand exactly why someone had done a particular experiment, what they had found out and not found out, and all the questions left over at the end. He was also a great storyteller and wit, and his classes were a delight. And since he was a bit of a rebel himself (he co-founded Gay McGill in the 1970s and survived, no mean trick for a young professor), he didn't mind all the bad things the chemists said about me. They had said a few bad things about him, too.

I started my M.Sc. work with him in 1967, just after my son Mikail was born and my husband (and, unfortunately, his mother) moved out. Life was a bit challenging. In that pre-feminist age, single mothers were frowned on, especially those with career ambitions. A visit to the pediatrician at the free clinic was a nightmare of humiliation – being interrogated as to whether my sons had a father figure in the house, being sent to see a social worker for counselling because my son's T-shirt had a stain on it.

In the summer of 1969, Fidel Castro was interested in cattle breeding and Professor Southin got a joint appointment to teach genetics at the University of Havana. He sold the University on the idea of my coming with him to help with teaching. It was the summer when all Cubans were summoned to the sugar cane plantations to harvest ten million tons of sugar, and the course never got organized. I didn't have much to do besides hang around the Habana Libre Hotel (formerly the Havana Hilton) with the other young people there.

Cuba attracted many Americans who came to learn about their new society. A group from an organization called Science for the People came to tour, and they were garrisoned at the Habana Libre. I met Len Radinsky, a tall, handsome physical anthropologist from the University of Chicago who was criticizing the idea that behaviour was genetically determined. A top-flight geneticist, Jim Shapiro, was also in the group. Shapiro was known for having figured out why some microbial genes turn on and off. He and several other geneticists were trying to figure out how they could use their scientific knowledge to transform society.[1] I was fascinated.

During the summer in Cuba, I thought over my choices. I was far from the Science for the People movement in the U.S., although I kept links with them after returning home. I had finally succeeded in arranging decent day care for my children in Montreal and didn't really want to leave. So I decided to grit my teeth and get my Ph.D. in genetics and then, when my children would be older and more mobile, I could find a way to link my science with my social ideas. John Southin was my model, since he was using his knowledge of genetics to help third-world countries.

I started a Ph.D. at McGill in molecular genetics of fungi with "Lynn," a young professor with children the age of mine. We got along well and I sympathized with her struggles in the department. Lynn had been hired at the same time as another young professor, "Harry." Both were experts in the same field and they worked together. Lynn was an excellent teacher and was assigned the big introductory genetics course, which took a lot of time and preparation. The students loved her and proposed her as Dean of Student Affairs, but she refused so that she could spend time on research. Harry described himself accurately as a terrible teacher. He was unable to explain anything, and he tended to get frustrated and yell and throw radioactive test tubes at the wall when a student made a mistake. He was therefore assigned to an advanced-level course with very few students. He got awful evaluations. At the end of their first three years, Lynn and Harry had each published three papers, two together and one separately for each. As far as I could tell, their contributions to research were equal and her teaching was infinitely better. When the department offered a promotion to Harry and a renewed contract at a lower level to Lynn, she was understandably annoyed and left. I was passed on to Harry.

My graduate studies took place in the late 1960s and early 1970s, when Quebec was dealing with the various ways that social justice, language, and nationalism could fit together. Students I knew demonstrated often and about all sorts of things. Even in the relatively staid biology department at McGill, professors and students were restive. I had another learning experience when the maintenance staff at McGill went on strike in the winter of 1973–74. With other students, I went out to support the picket lines with Harry and another young professor. They were showing courage because McGill professors were being fired for even the mildest political activity.[2] As a student, I wasn't in any danger of being fired and I had not thought much about the strike. I was out more as an automatic act of rebellion than anything else. But that changed when one of the picketing professors accidentally dropped a coffee cup on the ground in front of the biology building. I reflexively picked it up to put it in the trash, but a striking worker snatched it angrily out of the can and threw it back on the ground. In his eyes I had been scabbing, doing the job that the strikers were refusing to do. His body language showed me how serious the strike was for him. He explained to me how much he hated the way McGill treated the maintenance staff and made me understand that picketing wasn't just a fun thing to do while waiting for my cells to grow in the lab, but his only chance of getting respect from his employer. If I hadn't yet made it across the empathy gap, I was at least becoming more conscious of it.

I had other contacts with the world outside the laboratory as I engaged in the perpetual search for babysitting. Although my parents did what they could to help, my sons and I had little money, and at that time there was no government-supported day care.[3] In laboratory science, cells grow at unpredictable rates and procedures can abort or succeed in ways that change the time experiments take. The end of the workday is unpredictable, and I often had to leave before the cell cycle could be completed, slowing my progress and reminding my professors of how unreliable, unscientific, and disorganized I must be. After a few years of temporary solutions, I lucked out. I found a charitable day care centre, run and financially supported by women from the "best" part of Montreal. The teachers were good, the place was clean, bright, and well equipped, tuition was affordable, and the director, Alison, was a wonderful, warm woman with lots of great ideas about early childhood education. She understood parents' constraints and could deal with them. My son was happy at school and this was a load off my mind.

I became president of the parents' association and got to know a lot of the other mothers. They too were grateful to the nursery, especially since most of them were working at jobs that paid little better than my student stipend. For example, my friend Carol, another single mother, was finishing an accounting degree while working full time as a bookkeeper to support her three children, all under seven. The only apartment she had been able to rent was far from the day care, and at the end of her workday she had to race to pick up the children, make supper, and get the children ready for bed before going to her night classes. Since I too was quite familiar with the sprint at the end of the day, taking three successive buses to pick up both children and then climbing the hill to my apartment carrying the two-year-old and the bag of groceries, we got the idea of asking the nursery to provide a pickup and delivery service for the children. The nursery actually owned a bus, which was only used a couple of times a year for outings.

The parents' association was enthusiastic and we offered to organize and pay for the service if the nursery would lend us the bus and driver. Unfortunately, the director told us, the board would not allow it. Board members saw their role as educating the working mothers who used the day care so that they would accept their responsibilities toward their children. They thought it would be inappropriate to encourage the mothers to be even more "irresponsible" by doing yet another part of the mothers' job for them.

My mouth dropped open when I heard that Carol and I and the other mothers, who felt we were kicking wildly just to keep our families afloat, were being called irresponsible. Yes, some of us were occasionally a few minutes late picking up our children at the end of the day, but to us, every evening we were on time was a victory against the odds. Although I could understand that our benefactresses felt that they had given us enough and jibbed at adding in transportation, the fact that our efforts were invisible was a real shock.[4] It was during this struggle that I became fully conscious of the empathy gap between social classes, because I spent the day on one side of the gap as a graduate student but the minute I left the laboratory I was on the other side.

I finally got my Ph.D. in 1975. When my thesis defence was over and my professors congratulated me, I was relieved and happy. After five years, I was finally Dr. Messing. I had a few minutes before the party, so I floated on my cloud of happiness to John Southin's phone and called my parents to tell them the good news. I told the operator, "Tell them it's a collect call from Dr. Messing." The operator said, "I can't do that until Dr. Messing is actually on the line. Please call him to the phone." Bump! I was back in the real world.

I left Montreal that week for a U.S. research laboratory, but by then Montreal was home and I wanted to get a job there. While I was still in graduate school, Donna Mergler had asked me to help her teach a course on women in her biology department at UQAM. Donna's children were sick a lot that winter and I ended up teaching most of the course. At the end of my postdoctorate in 1976 Donna was able to get me a job in her department.

I loved the biology department at UQAM. Almost everyone was young and their research and approach to teaching was refreshing. French-language education in Quebec was just untangling itself from the church, and all kinds of experiments were beginning, among them the union-university agreement, a militant professors' union, collective governance of departments, and all kinds of pedagogical adventures. As a result, the students who chose our university were adventuresome and challenging, fun to teach. I remember in my first genetics course getting into an argument with Mimi's husband about whether my vision of the cell was "imperialist" because I taught that DNA determined cell physiology while for him there was an equal interaction between the nucleus and the cytoplasm.[5] Not a question that had been discussed much at McGill.

Viewed from the perspective of 2014 academia, my department was remarkably tolerant of the changes in my research program from fungus to humans and supportive of my relationship with the union-university agreement. After the call from the union in 1978, when I realized no one else was going to examine the workers' chromosomes, I had no choice but to do it myself. However, granting organizations were less adventurous than my department. Between 1985 and 1995, we got about $600,000 from various government agencies to examine the damaged genes of workers. However, the grants were for research on the type of test we used, not for studying workers' health. There was really no place that a union or a group of workers could ask for sufficient funds to find out whether they were affected by their working conditions.

So Donna Mergler and I were really happy when the Quebec government set up an occupational health research institute, called IRSST.[6] We were delighted to learn that it would be run with input from both unions and employers and it would generously support researchers that devoted themselves to occupational health. And, at first, the signs were all good. The people hired by IRSST came from all kinds of disciplines and backgrounds. Although the researchers who collaborated with the asbestos mine owners got money to try to show that asbestos from Quebec wasn't harmful, others were funded to identify new dangerous conditions. When IRSST announced its new teams program in 1983–84, Donna decided to set up a team to detect early signs of occupational health damage to genes (my specialty) and to the brain (Donna's). This was a wonderful opportunity to prevent health problems among workers and we were sure it was the beginning of a new age for workers' health.

We had a few hurdles to get over, though. It was 1982 and we were convinced that a team headed by women would never get a big grant from anywhere so we asked two of our male colleagues to join us. "Michel" was a well-spoken, forceful man who had been to the best schools and knew everyone. He was really good at impressing people and getting things done – the more opposition, the happier he was. As our previous department chair, he had succeeded in building successful programs and getting money to hire lots of professors. We thought he would be great as the team leader, despite his lack of experience in research. Another colleague, "Jacques," was a gentle man who had done a very successful occupational health study in collaboration with the CSN union. Everyone loved him and he was very much in demand as a speaker. The four of us got on well and, we thought, would make a wonderful team.

Donna and I, as the most experienced researchers, wrote a first draft of our team grant application and showed it to the guys. Jacques, as anticipated, was fine with everything and made no changes. Michel, on the other hand, hit the roof. What did we mean by saying that we would pay particular attention to women's occupational health? That would be unfair to men and the unions wouldn't accept it. We pointed out that we had been working with the union women's committees and that they thought it was important to include women in occupational health since women's work was often neglected. Hospital workers were exposed to disease, radiation, and chemicals. Women in factories had to do repetitive work at a fast pace. Sewing machine operators sniffed toxic dyes all day. But Michel, who had worked in union educational sessions with miners and forestry workers, wasn't impressed.

One beautiful spring day we went over to Michel's apartment to write the final draft. Jacques, who didn't like conflict, was busy that day. After we had spent an hour in his living room repeating our different positions, Michel took a stand. If we insisted on including the paragraph about women, he would leave the team. To our own surprise as well as his, Donna and I decided to let him go. We went down the steps and out of his apartment house into the warm sunshine. We started the first of several turns around his block, wondering how to replace Michel. We figured (correctly) that Jacques would not want to take sides against Michel and that he would leave, too. Who could we get to lead the team? We went through the names of all our colleagues, but after three times around the block we were getting discouraged. No one we knew with experience in occupational health had both leadership qualities and experience with workers. What about "Joseph"? Too dumb. What about "Sylvain"? He had always worked with employers alone. And "François"? No one could work with him.

Our legs were getting pretty tired when an idea occurred to us: could we lead our own team? Maybe if there were two of us, maybe together we could make the weight equivalent to a male head. We decided to go ahead, and were able to get funding for five years, although the head of IRSST was uncomfortable with the idea of our being co-heads. "Every train needs a locomotive!" he exclaimed. Donna and I kept straight faces but, after leaving his office, we chugged and hooted happily all the way down the street.

After five years, IRSST cancelled the teams program. Their board wanted to maintain closer control over its funding on a project by project basis. Research funded by IRSST became more closely linked to the priorities of the compensation system. Those of us interested in what they called derogatorily "preventive prevention," that is, prevention of occupational health problems before they became diagnosed diseases,[7] were out in the cold. The organization made a decision not to fund research specifically on women or gender.[8]

We were forced to go back to finding indirect ways to support research intended to protect workers' health. From 1990 to 1993, we were able to tap into small amounts of money from sources in the social sciences (see chapter 3). Then, from 1993 to 2004, the Québec Ministry of Health's fund for community-initiated research generously financed our collaboration with the union women's committees, ensuring that we could develop understanding of the problems of women workers (see preface). In the following year, the program was closed and we again resorted to scratching for funding from conventional sources. I retired in 2008 and my colleagues put in a last request for funding for our union-university partnership. We were refused because, said the committee, we published in scientific journals that were too applied – we were not at the forefront of theoretical advances. At the same time, Donna and I were unable to bring any of our trainees or young collaborators into our biology department. The department had moved on, and the ability to respond to community needs was not a priority. As I write this, a group of young professors is starting a new collaboration, hoping to respond to the community's occupational health needs, as they perceive them. But their chance of getting enough money to fund their program is much smaller than Donna and I experienced in 1982.

Now there are very few grants for scientists who are interested in collaborating with workers to do research aimed at protecting their health. First, there is a general chill on scientific research that is not directly related to making money for business.[9] This is especially true in Canada, but the U.S. is also suffering from vestiges of the Bush era. Second, many of the research questions that arise from worker-scientist collaborations are new or provide new angles on old questions. This used to be an advantage, because originality was a plus in a grant application, but now, with so little money to dispose of, many committees are reluctant to support what they call "fishing expeditions" or exploratory research. They are nervous about funding proposals that have unfamiliar ideas or seem to come from "out in left field." The committees are having trouble just finding money for tried and true research teams that use standard methods to provide reliable information about recognized problems. There is no more wiggle room to take a chance on a wild idea that might lead nowhere. Third, new research questions often require untested methods or critiques of existing methods. Since any competent journal editor, on reading a submission that criticizes "Jack Brown's" methods, will automatically send the submission to Jack Brown for review, it can be hard to publish the fruits of union-university collaboration.

This doesn't mean that no one is trying to address workers' occupational health problems. In fact, many creditable scientists empathize with workers, and several of them collaborate with unions or other community groups. They are working hard against the odds.

In the next three chapters, I will address the experiences and practices of "empathetic" and "non-empathetic" scientists. This division is a shorthand way of talking about the science produced by those who appear to me to understand the difficulties experienced by workers, versus those who do not. I will give examples of how science that is out of touch with workers' reality can end up using methods that don't detect true occupational health problems.

Chapter 9

Crabs, Pain, and Sceptical Scientists

I DON'T USUALLY CRY IN SCIENTIFIC MEETINGS. But in 2006, the annual meeting of the Canadian Association for Research on Work and Health in St. John's, Newfoundland made me cry, and I wasn't alone. On the last day of the meeting, I went to hear a collection of presentations about crab processing workers in Newfoundland and on the Lower North Shore of Quebec. Seafood catching and processing are just about the only regular paid work available to men and women in these remote locations. When the cod stocks collapsed in 1992 due to overfishing, the companies that had processed cod turned to crab and shrimp. The shellfish season is short – workers have to cram a lot of processing into about fifteen weeks. They have to process the shellfish as they come in so that it will be fresh for customers. This means they work many hours per day. Since the fishing boats come in every day so as to profit from the short season, the processors have to work many days in a row. They accept this because they need to earn enough money to carry them through the winter. Also, if they are not paid for enough total hours, they will be ineligible for unemployment insurance. They can't take days off even when they are sick because sick days won't count for unemployment insurance.

The women particularly suffer from work-related asthma due to exposure to crabshell dust and from musculoskeletal disorders due to intensely repetitive movements in a cold room.[1] Their pain can become very intense, worsening as the season progresses. Dr. Pierre Chrétien, a public health physician in the region, described to a roomful of researchers how he needed to work out giving the women just enough pain medication so they could continue to work without becoming addicted. Sometimes the pain was just too great for small doses and the women took more or combined medications and underwent withdrawal at the end of the season. His quotations from the desperate workers had us all in tears.[2] After his talk, I introduced him to Bob Sass from Saskatchewan, a pioneer in occupational health who had been working for years, politically and scientifically, to improve working conditions. Bob had just watched one of his good friends from the labour movement die agonizingly from the effects of asbestos exposure. Pierre and Bob shared hugs across the language barrier.

Both of these men, and many of the other scientists at the session on musculoskeletal disorders, were appalled by the health effects of the working conditions they were studying. Many of them have devoted their lives to trying to prevent suffering. No one could accuse them of not empathizing with the workers. And in fact there are a number of scientists who empathize with workers, understand their pain, and do their best to put an end to it. So an important question is: How do the institutions of science treat researchers who do and those who don't empathize? Let's start with those who don't appear to understand.

Some scientists are extremely sceptical about workers' pain. As researcher Bradley Evanoff put it in a letter sent to the Occupational and Environmental Medicine listserv (OEM-L), "this constant . . . denial of the whole concept of WRMSD [work-related musculoskeletal disorders] continues to amaze me. I know practitioners who will confidently diagnose lateral epicondylitis [a musculoskeletal disorder also known as tennis elbow] as the result of two hours of casual tennis, but who will not accept that 6 months of 50 hours per week of wire pulling and wire stripping [jobs that require very forceful and frequent wrist extension] can lead to the exact same musculoskeletal disorder."[3] In other words, many medical scientists suffer from an empathy gap. They play tennis and go to museums, so they can understand tennis elbows and museum fatigue. But they have gone to school for years and years so as to avoid doing repetitive physical labour at work – how can they sympathize with the problems of wire strippers? Often, they just don't believe in their stories.

Different sciences of how to treat workers' pain have developed in response to workers' claims for compensation for musculoskeletal disorders. One approach uses the concept of "fear-avoidance" developed by psychologists around fifty years ago to explain individual differences in pain-related behaviour[4] and later applied to work-related pain. Take the following research project, described in the Fall 2011 information bulletin of the IRSST,[5] Quebec's occupational health research institute. The researchers examined 202 people who had suffered work-related musculoskeletal disorders. The researchers first gave the workers psychometric tests that diagnose depression, "fear of movement," and "pain catastrophizing." The test for pain catastrophizing had been developed to test the idea that people who were scared and pessimistic about their pain would recover more slowly.

A year after their first test, the researchers re-examined the 202 workers with musculoskeletal disorders. They found that all three measures (depression, fear of movement, and pain catastrophizing) were related to how much pain the injured workers still felt and whether they had been able to return to work.[6] That is, the more depressed, fearful of movement and fearful of pain they had been at the start, the less likely they were to have recovered quickly. This was especially true among women. In their reports, the researchers expressed their conclusions as follows: "The findings of this study indicate [that] the depression impacts negatively on response to rehabilitation treatment and return to work outcomes"[7] and "pain catastrophizing and fear of movement act as differential predictors of long-term pain-related outcomes."[8]

How is this result to be interpreted? Reading the report and the article, we are invited to think that some injured workers, especially the women, were fearful and depressed. The more fearful/depressed they were, the more their pain persisted, and the less likely they were to be able to go back to their jobs after the injury. The article proposed a psychological intervention to change these attitudes.

But before sending the workers to the psychiatrist's couch, let's think about the test for catastrophizing and how it came to be. It turns out that it was developed and assessed with pain-free individuals, mostly undergraduate students. The students were asked to predict the pain they would feel in some experimental situations. Then experimenters inflicted pain and asked them to describe it. And yes, those more afraid of pain later described the experimental pain as more severe.[9]

Let me suggest another way to look at the same research results. The catastrophizing scale includes thirteen items from three subscales: Rumination ("I keep thinking about how much it hurts"); Magnification ("I become afraid that the pain may get worse"); and Helplessness ("There is nothing I can do to reduce the intensity of the pain").[10] Now, suppose we believe the workers when they say their pain is bad and their jobs put them at risk. This is a very different situation from a group of students who are not experiencing chronic pain.

Could the worker who is more badly injured and more exposed to more dangerous work be the one who is more likely to fear going back to the job? How does the catastrophizing scale work then? Perhaps the wire puller doesn't want to go back too soon because he knows pulling on the wires will hurt a lot. He worries about this: "keeps thinking about how much it hurts." And he is, indeed, "afraid the pain may get worse" and does think, "there is nothing I can do to reduce the intensity of the pain." So the catastrophizing scale may measure the risk of the job and the degree of pain, as perceived by the worker. If we consider that the workers, given their experience, might be good at evaluating this risk, it is not surprising that the higher the score on the catastrophizing scale, the more likely that the workers will have trouble when they get back to work, as they had predicted.

However, the approach that considers that workers are exaggerating their pain is well respected and well supported by public funds. They publish in good journals.

What about scientists who do empathize? There is a literature on how important it is for injured workers to be able to get their doctors to understand their point of view, and on the validity of workers' perceptions of pain.[11] Are those scientists respected and do they get funds? Sometimes. Dr. Michel Vézina, a public health scientist and physician from Laval University in Quebec, has devoted his career to identifying and preventing damage to workers' health. From his position at the Québec National Institute of Public Health (INSPQ), he lobbied for years to get a regular survey of workers' health in order to inform and orient prevention efforts. He finally succeeded. In 2002, the Quebec government passed a law saying that workers' health would be surveyed every five years.

The first study, called EQCOTESST,[12] took five years to set up. A committee was set up to oversee it, with members of the INSPQ, the Ministry of Labour, the Labour Standards Board, the IRSST, and the Québec Statistics Institute (ISQ). In 2007–2008, professional interviewers questioned over five thousand workers. The entire team of eleven principal researchers scrutinized each question, and each of the nine subject areas profited from additional collaborators with specific expertise; all eighteen of these people participated in data analysis.[13] When the report was eventually submitted to the overseeing committee, two or three outside evaluators assessed the scientific quality of each chapter and some sections were rewritten or nuanced in response to evaluators' questions. In the summer of 2010, the committee sent the final revised, reviewed, and approved report to the IRSST. It disappeared for over a year.

No one has told me what happened to the report during the lost months or why it was suddenly released late in the day on September 20, 2011. I do know that, no later than 10:58 the next morning, it was denounced in no uncertain terms by the three major employer associations in Quebec.[14] They "raised serious questions about its scientific value"[15] – scientific value ensured by the supervision of IRSST, an organization they fund and control jointly with worker representatives, not to mention the neutral government organizations ISQ and INSPQ. During the following days, the employers could be heard on the radio and in the newspapers, reiterating that the study results should not be believed – working conditions in Quebec were excellent. Union counsellors told us that, in the workplace, employers were telling health and safety representatives not to listen to the results, that the study was flawed.

A few weeks later, I had lunch with an IRSST staff member who told me IRSST was in trouble. Its management was afraid it was going to be spending the next year defending EQCOTESST and would have no time to do anything else, especially anything controversial. Knowing I wanted to ask for money to study prolonged standing, my informant wanted to warn me that this was not a good time. A few weeks later, one of IRSST's most respected researchers resigned. The rumour was that the researcher had been forbidden to defend EQCOTESST. The report, which had cost over a million dollars,[16] had been orphaned by its parent organization.

I next heard about EQCOTESST in public at a colloquium on occupational health and safety held in Quebec City. A representative of the employers' association denounced the report saying that it was based on "people saying that they had felt pain at some time during the previous twelve months." But, he objected, "Everyone feels pain at some time." Luckily, I had my computer with me and was able immediately to read out the exact question. It asked not about just any old pain but about "significant pain that interfered with usual activities often or all the time" over the previous twelve months – hardly a commonplace situation. As a scientist, I would have been totally embarrassed by being shown up like that, but the employer representative was not daunted. In fact, six weeks later a similar thing happened in Montreal. The same man and one of his colleagues were at another colloquium and tried again to make the case that the question on pain was too broad. Again, one of the scientists present read out the true question and, again, the employer representatives were forced to sit down. I expressed my surprise at their persistence to a union representative, who said, "That's what they do. They say the same lie over and over in the hope that it will become the truth. And it's working."

Why? How could the employers' organizations, with no particular qualifications as researchers, seriously threaten the EQCOTESST study, which had been backed by all the institutions in Quebec with occupational health expertise? How could they raise such a fuss that the research institute was paralyzed? How did they manage to ensure a media silence around a study that offered information on all Quebec workers?

The employers' argument was that the study was not scientific because it listened to workers: "The study is essentially a survey of perceptions carried out only among workers and not an analysis of objective data and established fact."[17] They were referring to the fact that the questionnaire asked workers about their working conditions and their health problems, rather than getting expert opinion on working conditions and asking a physician about their health problems.

But the EQCOTESST scientists had already studied the validity of self-reports and knew quite well what workers could report accurately and which conditions needed to be observed by scientists. There is an extensive scientific literature on this.[18] We know, for example, that workers can correctly report whether they usually stand at work and whether they have access to a chair, but that they often overestimate time spent standing without moving.[19] We know how well workers estimate their frequency of repetitive movements and intensity of physical effort. We know also that even the best ergonomists only sample specific times, usually very short times, during one or two weeks, whereas workers are at the job forty hours per week over the whole year.[20] A serious case can be made that it is more scientific to ask workers about some exposures than to ask scientists – the workers actually see the working conditions all the time whereas the scientists must choose defined, limited times to sample.

Surprisingly, we don't know as much about the accuracy of information that comes from scientists. Few people have studied what scientists report accurately and what they don't. One set of disquieting results has come from a team of Greek scientists. Dr. Ioannidis and colleagues checked the medical literature and reported that most published research results could not be replicated.[21] And we have no information on the accuracy of employer reports on working conditions.

For some reason, it is nevertheless considered scientific to come in and sample workplace temperatures, air quality, or working postures at specific sites for a few hours and then generalize about them to the working environment, without checking with the workers to see whether the sampled period is typical of the work week, whether the sites sampled are representative of all sites, whether working conditions have recently changed, or whether the list of sampled conditions is complete. On the other hand, many people think it is unscientific to ask the worker who has been in the workplace all year to report on environmental quality. "Rigour" and "objectivity" seem to be in the eye of the beholder.

As a consequence, perhaps, empathy doesn't seem to be a highly valued attribute of scientists. Scientists are not taught to listen to workers – in fact we could say scientists are taught not to listen to workers. Scientists who overtly sympathize with workers are swimming upstream. One consequence is that these "empathetic" scientists have less leeway than others to explore new methods. The burden of proof is on them to show that they have been careful, objective, cautious, rigorous, and conventional. They have to prove every step of their reasoning by reference to existing scientific methods and results, preferably quantitative methods and results published in reputable, peer-reviewed publications. By a kind of circular argument, the most reputable journals are not generally those with union connections, but can be those with employer connections.

This situation is unfortunate because contact with workers can give scientists ideas that can lead to new methods and important results. But worker-inspired ideas can be a very hard sell. In 1987, we spent a day presenting our research to a peer review committee that eventually gave our group half a million dollars. The committee grilled us about every aspect of our research methods. They were nervous about giving us money and particularly wanted to be told that we were going to use standard laboratory tests. It was understandable – we were young and female, and they knew we were associated with unions. I almost blew the whole thing when I presented my ideas about statistics, ideas that had occurred to me while exchanging with radiation-exposed workers (see the next chapter). One committee member summed up his reservations saying, "But when you submit your articles to scientific journals, you're going to put the risk at .05, right?" I am sorry to say it took me less than a second to reassure them that I wasn't some kind of a statistical revolutionary. Everyone in the room went back to normal breathing and they eventually gave us lots of money. It was more than twenty-five years before I was able to find a context where I could publish the idea in the scientific literature.

After talking with workers and becoming aware of the details of their worksites, Donna and I also had the idea of doing gender-sensitive analyses of workplace risks. We had met a lot of people, employers, union representatives, and doctors, who thought that women in general complained more than men about their working conditions. Scientists also appeared to show that women complained more about the same conditions. The image of women complaining about nothing corresponded to some people's stereotypes and was not rigorously critiqued in the scientific literature. By this time, Donna and I had spent enough time watching workers to know where the women and men worked. In Donna's studies of poultry processing, men were at different points of the production process from women. So the same name of a workplace condition (say, "exposed to cold temperatures") didn't mean the same exposure in a woman's job as in a man's job. The women exposed to "cold" temperatures usually stood all day in the same place at 4°C to 10°C, while the men exposed to "cold" temperatures usually walked around from one area to another, with temperatures ranging from -10°C to +10°C. Similarly, male food servers take many fewer steps at work than females in the same restaurants so "walking" means more impacts on the knees and feet for women than for men.[22] On the other hand, "lifting weights" may involve lighter weights for women, involving less stress on their musculoskeletal system – but this is countered by the fact that the weights lifted by women are more often people (who wiggle and resist) while those lifted by men are usually objects.

Because of women's and men's different working conditions, we suggested that researchers studying working conditions consider them separately. Researchers at CINBIOSE who developed this new idea worked together to examine the effects of this gender-based segregation of the labour market on workers' health.[23] Academics were slow to adopt this approach. A union-based organization found the approach interesting and publicized it in Europe[24] but it took almost thirty years for this idea to be accepted by enough scientists to be presented as mainstream science at an international occupational health conference.

A final example of the fate of "empathetic" science comes from the controversy around psychological symptoms and musculoskeletal disorders. A lot of studies have found statistical associations between depression or psychological distress and work-related pain in the back, neck, shoulders, and arms. That is, workers with pain are often those who are depressed or in distress. And the same working conditions, such as repetitive movements, can be associated with the physical and the psychological symptoms. People who work fast and who have little leeway to change their posture or their methods tend to suffer more in every way, physically and psychologically. When scientists first noticed a link between psychological distress and musculoskeletal symptoms, my colleagues and I were not surprised. Of course people in pain would feel distress. But we quickly became aware that not all scientists interpreted this result the same way we did. Many were convinced, on the contrary, that the musculoskeletal symptoms were *caused by* the psychological distress, that people who were unhappy would therefore experience pain that they would, mistakenly, attribute to their work.

Which comes first in time, the distress, or the pain? Some scientists studied workers at intervals, starting with pain-free workers.[25] They found that the psychological symptoms tend to occur before the physical symptoms, which to them clinched the argument that physical symptoms result from the psychological symptoms. This research result generated a lot of interest and probably made a lot of money for consultants on stress reduction in the workplace. But wait a minute. The fact that repetitive work or work in awkward positions makes me bored and unhappy before it gives me tendinitis may not prove that the boredom and depression *cause* the tendinitis. It may just be that doing unpleasant work quickly results in boredom and bad moods, but that actual pain requires chronic inflammation and takes a bit longer.

With these results, it seems to me that it would be most reasonable to ask workers with physical pain to describe how it came about; how do they perceive the relation between their physical and psychological symptoms? I am afraid it would be hard to get funding for such a study. And the results would probably be called "subjective," as of course they would be. But no more so than asking a doctor or the employer.

The solution to the funding problem would be to have a grants program that included community groups such as unions in a decision-making capacity. In Canada, we had several such programs in the 1990s, but they have been terminated. There has been a proliferation of "community partnership" programs, but the new ones are better suited to programs that involve large private companies and ill adapted to impecunious community groups that can't participate financially.

Chapter 10

A Statistician's Toes and the Empathy Gap in Scientific Articles

LAST APRIL PIERRE AND I WERE CYCLING in Peru and learned that Peruvian dogs hate cyclists even more than North American dogs do. As Pierre was pedalling up a hill between Urcos and Combapata, a beige dog that had been lying peacefully by the side of the road with two little girls bounded up and bit him in the leg. A nice deep bite with jagged edges, dripping blood. The two girls immediately ran off with their dog along a narrow dirt track down the green valley, leaving us with a mathematical problem. What were the chances that Pierre would get rabies?

Rabies is almost 100 per cent fatal, but symptoms don't show up for some time, and victims who get vaccinated in the meantime can be saved. In Peru, the procedure consists of seven shots, injected into the abdomen on seven successive days. The vaccination is given whenever the dog can't be identified and tested for rabies, which was our case. Pierre's reasoning went like this: the dog didn't look like it had rabies; Peru has a universal, well-enforced anti-rabies vaccination program for dogs; the dog looked well cared for; we're on vacation, we don't want to be stuck for seven days in Cusco; therefore there is a zero chance that I have to get these shots. It all sounded very reasonable to me until the middle of the night, when I burst into tears. Yes, I thought the chance of Pierre's getting rabies was probably less than one in a thousand, but that was an unacceptably big chance because of those pesky millicupids. So Pierre got the shots. He didn't get rabies. The Peruvian public health system paid for almost everything and our sentence to spend a week in Cusco – in the heart of the Andes – turned out not to be so bad.

What is an acceptable risk? Risk to whom? Risk of what? Who decides? And who pays? In this case, two well-off Canadians were helped by a relatively poor nation to combat a risk that was probably less than one in a thousand ($p<0.001$). In other words, there was a 99.9 per cent chance that the intervention was unnecessary. Is this the general practice in health statistics?

Not really. Technically speaking, a risk is considered to be associated with a health effect if the chance of being wrong in making that association is less than 0.05. In other words, in occupational health, prevention is accepted when there is a 95 per cent chance that it is necessary, not 1 per cent. How was this 0.05 level chosen? The story goes that R.A. Fisher, the father of statistical analysis and statistical testing, was asked when a difference between two sets of numbers should be considered statistically significant. He went home to think about it, and got into the bathtub. While scrubbing between the toes of his right foot, he said to himself, "Five . . . five seems about right." Thus, we are told, the critical level for rejecting a hypothesis was set at 0.05 in the early 1900s.[1] And, ever since, Fisher's toes have determined when a drug is considered effective, which germs cause which diseases, and which working conditions are dangerous. If Fisher had scrubbed both feet, occupational health practice would be very different and a lot more workers' compensation would be paid out. If he had been scrubbing the toes of his pet two-toed sloth, life for injured workers would be even harder than it is today.

Why do I say this? The explanation is a bit complicated, but here it is. Fisher's suggestion was that two groups should be considered to be significantly different if the chance they are the same is less than one in 20 (<0.05). Lets say that a new cancer drug is being tested on half of a group of 100 patients. The drug is considered effective if the number of people (say 19) who get better among the 50 in the drug-treated group is significantly larger than the number of people (say 10) who get better in the untreated group. The statistician needs to calculate whether there is less than one chance in 20 that the cure rates are the same for the two groups. If the cure rates are not significantly different, the drug is considered to be ineffective. In this case, the standard test, known as "Fisher's exact test," says the drug company needs to work more on its product; despite appearances, after doing the math, 19/50 is not statistically different from 10/50 at the 0.05 level. The 0.05 level test says that the 9 extra cured patients could have gotten better by chance. If Fisher had looked at both his feet and set the level of significance at one in ten (0.10), the drug would have been approved. The level of 0.10 is a less stringent standard than 0.05.

The problem is, if 0.10 were to be used as the critical significance level, we would be less sure that the drug really worked. More of the drugs approved with the less stringent standard would eventually prove ineffective. So, to avoid the danger of putting patients on useless drugs, health scientists have been happy to accept the 0.05 level of significance. It seems like a reasonable standard to them. Pharmaceutical companies might prefer using 0.10, but then again, they might get sued more often, so they haven't protested too much. So this is why the 0.05 significance level for saying two groups are different (called "alpha") is used throughout the scientific literature in public health to make decisions about the relationship of drugs to cures, of diseases to causes, in fact about all associations between environmental components and human health.

I should point out that we also have to worry about the fact that putting the critical level of alpha too low might cause us to reject some really good cancer drugs. "Beta" is the name for the opposite problem: the probability that we will miss a good drug because, by chance, our drug trial rejected it. Beta has been arbitrarily set at 0.20 in many textbooks. But note that this is four times the number of toes used for alpha. Why the difference? Probably a practical one. For technical reasons, putting beta at 0.05 would make it necessary to study much larger samples of patients and it would cost a lot more to test drugs.

Putting alpha at 0.05 and beta at 0.20 has become standard practice in public health research and has therefore been applied to decisions about what is dangerous for workers. Is this an appropriate application of statistics? What are the consequences of these technical decisions on alpha and beta for workers' lives? Let's take a study of the relationship between back pain and lifting weights. In order to suggest limits on what weights should be lifted at work, scientists have done many studies of workers' capacities.[2] However, for ethical as well as practical reasons, no one has systematically exposed workers to increasing weights in order to determine how many workers have back disorders after lifting how much weight for how long.

Instead, scientists have studied "convenience samples," groups of people who lift weights at work, to see whether their level of back pain differs significantly from those who don't lift weights. If a group of workers lifts about 30 kg repeatedly all day, the number with back pain will be compared to that of another group of workers who don't lift weights regularly. If there is less than 1 chance in 20 (alpha of 0.05) that their levels of pain are the same, scientists will decide that 30 kg is OK. If Fisher had looked at both feet in his bath (alpha of 0.10), more groups of workers would be considered to have back pain related to their work. More workers would probably receive workers' compensation for their back pain, and the weights lifted at work would probably be lighter for everyone. The use of the 0.05 level determines that workers' health is only protected if there is less than one chance in 20 that a particular working condition causes health damage.

What about beta in this example? Beta is the probability that we will say 30 kg is OK when in fact lifting 30 kg does really cause back pain. For mathematical reasons, in order to keep beta low, we need to study a very large sample of workers who regularly lift about 30 kg. This kind of sample is hard to find, because employers don't often consent to have ergonomists come into their workplaces to see what their workers are exposed to. So, in practice, a lot of studies are inconclusive by scientific standards. These studies are reported in the scientific literature as saying that there is no proof that lifting 30 kg is harmful. So a lot of workers can have aching backs before science will lower the weight limit.

What if the workers or their families got to decide on the level of alpha? It might go a lot higher. If I got to decide on the science that determined whether Pierre should lift 30 kg all day and risk chronic back pain, I would want that level to be set at 0.999, just as I did in Peru. Would that be better?

Maybe the answer lies in an ordinary court of law, where decisions have to be made one way or another without dilly-dallying. In court, the decisions are made on the "weight of the evidence."[3] That is, the court would decide if it were more likely than not that the workers would have back pain while exposed to heavy lifting. In this case, a level of 0.50 would be used, ten times higher than the scientists' level of alpha. A court would easily find a link between back pain and heavy lifting.

Professor Katherine Lippel, a CINBIOSE researcher and legal specialist, has thought a lot about how scientists perceive and represent evidence and how tribunals judge evidence in workers' compensation. She was very concerned, in 1994, when a tribunal refused compensation to three women who sorted mail for the Canadian postal service on the grounds that their work had not caused the pain in their arms.[4] Nicole Vézina, asked to examine the job, had found that the postal workers picked up packages of letters weighing on average just under a kilogram and typed in the postal codes (pressure of 71g per keystroke, 7920 keystrokes per hour). They processed about twenty of the packages per *minute*, one by one, every day. The three women had severe chronic pain at the elbow joint and claimed workers' compensation, saying that the pain was due to the repetitive package lifting they did on the job. Since the tribunal based its decision on the scientific literature, Katherine asked me and two other health scientists to look at the written decision.[5]

You could have knocked me over with a feather when I read that the tribunal had based its refusal to compensate on the language in an article by Dr. Barbara Silverstein. Barbara, an ergonomics researcher with the Washington State Department of Labor and Industries, is one of my role models. In the 1980s, she was one of the first scientists to link repetitive motion in factories to musculoskeletal disorders like carpal tunnel syndrome and tendinitis. In the 1990s she was the architect of the short-lived U.S. ergonomics regulations, designed to reduce musculoskeletal disorders by reducing the need for force and repetitive movements in the workplace. These regulations were promulgated in the last days of the Clinton (Democratic) administration but repealed immediately by the Bush (Republican) administration after protests by employers. In the early years of this century, excluded from federal policymaking, Barbara returned to Washington State where she designed its ergonomics regulations. Her regulations were put into force in 2000 but, like their federal counterpart, they were repealed three years later under pressure from industry. She is continuing to do important research on musculoskeletal health and working toward reinstatement of the regulations.[6]

Because of her commitment to preventing musculoskeletal disorders I would never have expected that a tribunal would cite any article by Barbara as evidence *against* compensation for the postal workers. Here's how it happened. Susan Stock, a physician and scientist specializing in musculoskeletal disorders, testified before the tribunal that Silverstein and her colleagues had published the best evidence that the type of movement done by the post office workers was associated with upper limb musculoskeletal disorders (also known as "cumulative trauma disorders"). Susan said Silverstein's articles convinced her that repetitive movements were risky for the workers' health. She concluded that Barbara had produced evidence that repetitive movements were risky, and that workers exposed to repetitive movements and suffering from cumulative trauma disorders should be compensated. But Judge Elaine Harvey disagreed. After reading Barbara's article, she rejected the application for compensation, saying, "Silverstein herself, whose research constitutes the basis for the arguments of the workers' expert [Dr. Susan Stock], stops well short of concluding [*se garde bien de conclure*] that her work has established a causal link between repetitive work and the appearance of cumulative trauma disorders."[7] Harvey's judgment was publicized by employers and became a precedent; during the two following years, the rate of accepted claims for compensation for similar disorders dropped significantly, from 46 per cent to 32 per cent.[8]

How could this have happened? How could anyone think Barbara Silverstein, a tireless worker for ergonomics regulations, didn't believe that repetitive movements caused musculoskeletal disorders?

What had Barbara actually said in her article? After presenting her results, she had discussed several aspects of her research design that might limit the interpretation of those results. Some of her scientific choices could have made the association between repetitive or forceful movements and musculoskeletal problems look stronger than they were in fact, and some could have made them look weaker. She then concluded, "Our findings *may* help in directing workplace interventions in the worker exposure disease cycle because they *suggest* a strategy for primary prevention. Through job modification a reduction in force or repetitiveness *may* result in a reduction in the prevalence of CTDs [cumulative trauma disorders]."[9] In other words, Barbara was only certain at 0.05, not completely. It was this type of hesitant language that allowed Judge Harvey to justify her belief that Dr. Silverstein didn't really believe repetitive movements were related to CTDs. But in fact, 0.05 should have been amply sufficient for a court decision. As Katherine Lippel pointed out, 0.49 should be sufficient for the purposes of a court decision.

In his book *Doubt is Their Product*, a careful chronicle of the misuse of science by corporations with products to defend, David Michaels, an epidemiologist, points out that rigorous scientists can never be entirely sure of anything, since there is always the possibility of an alternative hypothesis, new data may appear, or we may think of flaws in the data.[10] Scientists are trained never to make bald, unqualified statements. I learned at school that I should never express certainty. In fact, the scientists who evaluated one of my early papers scolded me for putting a 0 probability in one of my tables. Certainty at the level 0.000 does not exist, they growled. I had to say "<0.001."

There is a convention in peer-reviewed journals that the tone of articles should be reserved, to reflect that uncertainty. If Barbara had used language like "Our findings *should* direct workplace interventions in the worker exposure disease cycle through job modification. A reduction in force or repetitiveness *will* reduce the prevalence of CTDs," she might not have made it through peer review. Even to me, after forty-five years as a researcher, that unreserved last sentence sounds to me a bit like pamphleteering – out of place in science. But I am sure that Barbara, who loves kayaking, would not have wasted her time working weekends to try to reduce workplace exposures to repetitive movements if she didn't think they were harmful. It is a question of the literary style of academic publications, not the substance.

It is hard to believe that that Judge Harvey hadn't figured out that Silverstein's language was only a scientific convention. In fact, it is hard to believe that *anyone* could think that the elbow pain of women who had worked year after year moving one-kilogram packages at a rate of twenty per minute – that's over eleven thousand kilograms a day – was unrelated to their job. Especially given that there were three women with disabling elbow pain at that job.

Doctoral student Stephanie Premji, Katherine Lippel, and I decided to look into this question of evidence a bit further.[11] We first examined the language of scientists in twenty articles that examined the relationships between musculoskeletal disorders and workplace conditions, to see how they would talk about the relationships they found. We chose the articles from two journals, one (*Journal of Occupational and Environmental Medicine*) more closely associated with employers and employer funding than the other (*American Journal of Industrial Medicine*).[12] We predicted that articles in the journal more closely associated with employers would be slower to associate workplace conditions with musculoskeletal disorders.

That was not what we found at all. We found that scientists used cautious language in both journals. The hesitant language surprised Stephanie, who had been trained in qualitative research, but to me it looked normal.

What did surprise me was our finding that almost all authors, in both journals, used more caution and more statistical finesse in expressing positive results (showing a connection between work and health) than negative ones. For example, one group of researchers found that people exposed to repetitive bending at work were over seven times more likely to suffer from low back pain requiring sick leave of more than a week. But the authors didn't say, for example, "Work stations and jobs should be designed so people aren't forced to bend." Instead, they concluded in these terms: "the implications . . . are that both the level of biomechanical exposure and the psychosocial work environment, especially social support, *represent important dimensions to consider* in the reduction of work absenteeism."[13] In this article, when authors reported that a workplace condition was associated with poor musculoskeletal health, they reported the exact strength of the association, their level of uncertainty with the association, and their chance of being wrong (alpha) if they said there was an association. But, on the other hand, when they found an association to be too weak to be statistically significant, they said right out that there was no association. Their language was much less ambiguous, they did not usually report the strength of the association or the level of certainty of their results, and only three of the twenty studies even reported beta, their chance of being wrong in saying there was no association. I have to add that my own publications are not exceptions – almost everyone follows this procedure.

Let me point out the importance of this double standard for workers. What it means is that before a scientist will say that a working condition *may be* dangerous s/he has to be very, very sure that the working condition is associated with the danger to health. S/he has to have less than a 5 per cent chance of being wrong. If the scientist has a 10 per cent chance of being wrong – still a pretty small chance – by convention s/he is supposed to say that there is no link between the condition and health – even if the chance of being wrong in saying there is no link is much greater, like 20 per cent. So the cards are stacked against the workers.

What's more, we found no difference in this regard between the employer-friendly journal and the more neutral journal. Both sources were more critical toward positive results. So the caution in reporting positive associations doesn't seem to be directly motivated by a desire to please employers. It is the informal rules for scientific expression themselves that are hurting the cause of workers.

We did suspect that the influence and power of employers might have influenced the rules for scientific expression, though. In occupational health, there is always the risk of a court case. As I pointed out in chapter 1, scientists do not enjoy being cross-examined and positive results are more likely to get you into trouble with expensive lawyers lined up against you.

Stephanie and I concluded our study with two suggestions for researchers: (1) more balanced treatment for positive and negative results and (2) more forthright language so that judges and other lay people would know what actions to take. When we published the study, some of our friends got quite annoyed with us. Worker-friendly scientists objected to our suggestion that scientists be more forthright. They were understandably concerned that scientists who made unambiguous statements would be opening themselves up to attack.[14]

Thus, empathetic scientists have learned to disguise their wish to improve working conditions. The result is that there is a bias in what is published in the scientific literature, and that bias hurts workers. In part because of issues around proving causality, most claims for compensation of musculoskeletal disorders are refused.[15] And claimants are only the tip of the iceberg. For each claimant for musculoskeletal disorders, there are four workers with absences due to musculoskeletal disorders who don't claim, even though they think their disorders are work-related.[16]

Why don't they claim? Mostly because the process is so painful. Katherine Lippel (the law professor who studied the postal workers) has interviewed eighty-five workers in the process of claiming for work-related injuries and illnesses.[17] More than half the interviewees felt they weren't believed when they said their work was making them sick: "They think that we make things up." In fact, the rules of the game are set up so as to encourage scepticism. Workers have to go 95 per cent of the way toward persuading the health and safety professionals that their work has caused their health problem; scientific evidence is hard to get. In fact, Katherine concluded that the claim process itself could be responsible for injuring workers' health, because of the stress involved. It seems the whole system of occupational health, from research to compensation, has evolved to preclude empathy with workers.

Chapter 11

Can Scientists Care?

WHILE I WAS WRITING THIS BOOK, I showed the beginning chapters to a group of bright graduate students being trained in political economy by Professor Pat Armstrong at York University. They liked the stories, they wanted to know all about ergonomic analysis, but they weren't sure about the notion of "empathy gap." They thought the expression sounded as if it referred to an accidental breach in communications. But according to the students, if people don't listen to workers, it's probably because they don't want to.

And in fact, I had wondered whether that might be the case. After supervising the women with the coloured wires for over ten years, it was strange that my father hadn't figured out that they were intelligent – and bored. But did he really want to know this? Wouldn't it have made him feel uncomfortable, since he was a nice man and he had to supervise them doing an uninteresting job?

I'm not a political economist and I analyze work, not social movements. But my own experience tells me that it takes more than niceness to be able to empathize. Empathizing can come with very heavy obligations. When I explained to M. Lejeune, the head of human resources at Qualiprix, that his changes in policy wouldn't be much help to the women with family responsibilities, he told me firmly that he wasn't interested in helping them. In fact, he said, "I don't want to hear about work and family ever again." His job was to reduce turnover among his current employees, not to create jobs that would attract people he had never seen. And the people he wanted to retain were skilled personnel, to whom he referred in the masculine gender. The women on his staff who had family responsibilities were mostly cashiers, with no recognized skills, so why should he put time and money into improving their schedules? And into adapting the job so they could hire even more women with family responsibilities? Not on his radar.

I could get mad at M. Lejeune, but, in similar circumstances, my own behaviour hasn't been very different from his. When I was a new professor in the early 1980s and my students wanted to make our laboratory free of toxic fumes, I wasn't inclined to listen. My job was to produce publishable results or my grants wouldn't be renewed. My performance wasn't going to be evaluated based on whether my students felt dizzy or uncomfortable or even whether they thought I was a good supervisor. I was going to get future research funds based on how fast my students and I finished our experiments and wrote up the findings, and we had better get on with it. Hadn't I pipetted stinky toxic chemicals by mouth during my own graduate studies and wasn't I OK?[1] Besides, none of the other professors were worried about the smells on our floor.

In short, I wasn't feeling at all empathetic, I was feeling annoyed. If Micheline, Ana María, and the other students hadn't made it clear to me that their problem wasn't going to go away, I would probably have done nothing, and in fact I didn't do much. It was only once someone in maintenance had fixed up the ventilation that I could feel empathy for the student who was worried about exposing her fetus to the fumes.

How did the problem get solved? Basically because my students were persistent and insisted on being heard. They contradicted their employer and kept on doing it. They insisted that I and the other professors listen to the fears of the pregnant student. And they created structures that made empathy possible and desirable. Now, thirty years later, thanks to those stubborn students and people like them, health and safety training, industrial hygiene procedures, and thoughtful waste disposal are standard at UQAM, and all researchers are subject to pretty much the same rules. And our student research assistants have their own union to watch over them.

Unions are confronted all the time by similar situations. My union collaborators are in the business of making employers listen. But changing working conditions doesn't come easily. Someone, somewhere has to be stubborn. The workers and their unions need to be courageous and persistent. It helps if workers can get support from scientists. For example, there is a model in the way women workers have succeeded in getting and keeping "precautionary leave" for pregnant and nursing workers in Quebec despite heavy and recurring opposition. The leave was included in the 1979 occupational health and safety law in response to pressure from feminists in the trade unions. The proposal was supported by letters in the press and other pressures from a number of university researchers and medical specialists. The law as eventually passed provided that pregnant and nursing women exposed to conditions dangerous for the fetus or nursing infant could ask to be reassigned to less hazardous jobs if such jobs were available. Failing reassignment, they could apply to go on leave until the end of the danger, or the end of the pregnancy/nursing – whichever came first. This leave is in addition to the standard maternity and parental leave, which at this writing lasts a year and can be shared by both parents.

According to Romaine Malenfant who has studied this, at first no one thought the program would affect many women, so it passed without objections. After all, not that many women of childbearing age were working at that time, and their jobs didn't look at all dangerous.[2] It took two years for the Occupational Health and Safety Commission to initiate a program to inform workers and employers about precautionary leave. Even before that, the unions started sponsoring educational sessions where women came to ask specific questions about their exposures at work so they would know when to ask for leave. In fact, scientists already knew that a number of very common working conditions (like prolonged standing) were bad for pregnant women and their fetuses. The 1980s were a time of intense scientific interest in occupational health, and researchers identified other conditions dangerous for pregnancy such as shift work and many of the chemical and biological exposures in hospital work. The demand for precautionary leave rose sharply. Since most employers had no idea where to reassign pregnant women – there are just not that many jobs in hospitals where health care aides will not be exposed to infection, for example – many women were sent home for the duration of the pregnancy. Although those who had introduced precautionary leave were primarily thinking about protecting women from chemicals and radiation during the first few months of pregnancy, a lot (60 per cent) of precautionary leaves in fact began in late pregnancy because of exposures to poor biomechanical conditions like cramped postures. About 40 per cent of pregnant workers now access leave, and only about 5 per cent of claims are refused.[3]

Over time, the Occupational Health and Safety Commission and many employers began to complain that the program cost too much, although it only represents about 6 per cent of the workers' compensation budget.

About every five years during the 1980s and 1990s, there were court challenges and attempts to make claiming more difficult or to transfer the program away from the occupational health and safety regime. In 1988, 1992, and 1998 the press carried denunciations of how pregnant women were exaggerating and how something had to be done to stop the "abuse," but the threat of danger to pregnant women and of the risk of malformed children is very powerful. Each time there was an attack, an informal coalition formed among union women's committees, front line health care workers, and scientists. Meetings were well co-ordinated and efficient and got the word out fast. Research results showing the dangers of some working conditions for pregnant women were highlighted. Researchers like Romaine Malenfant and Marc Renaud[4] demonstrated that the program was being used appropriately. The media were contacted and the attackers were pushed back.

At the same time, economic pressure increased to make the program more focused on improving the workplace conditions for pregnant women and less toward sending them home.[5] The proportion of women being reassigned (as opposed to being given leave) went from 15 per cent to 40 per cent. This was a good thing, because applications for precautionary reassignment could end up helping all workers. For example, in some credit unions, when pregnant women were given seats, all tellers in the branch were then allowed to sit down. In some hairdressing salons where workers were exposed to dangerous chemicals, new ventilation was installed that benefited everyone. Also, because of the activity around precautionary leave, the public began to realize that women's jobs weren't as safe and easy as they looked. Precautionary leave became an accepted part of pregnant women's lives in Quebec.

More recently, data have become available on the success of the precautionary leave program in protecting health. Dr. Agathe Croteau and her colleagues compared women who were exposed to specific dangers at work and had accessed the leave to women who were exposed to the same dangers and stayed on the job. The risk of preterm birth and low birth weight dropped significantly for women who went on precautionary leave before the twenty-fourth week of pregnancy.[6] The results showed that the precautionary leave was doing its job.

But the employers' defence organization Conseil du patronat du Québec, still hoping to make big changes in the health and safety legislation, immediately hired another scientist to discredit Dr. Croteau's findings and produced a report saying her study was unscientific.[7] The employers then called for the transfer of precautionary leave away from the purview of health and safety authorities, doing away with the impetus to improve the workplace. They announced their intentions to make changes in the law in December 2010, just as everyone was going home for the holidays.

Their strategy did not work, thanks in great part to Dr. Robert Plante. Robert, a public health physician and co-author of the Croteau study, had been a defender of precautionary leave from the beginning. For twenty years, he had been supporting of women who applied for leave. He had been a major player in all the coalitions in favour of maintaining the program.

That December, Robert was being treated for the stomach cancer that would kill him eight months later. He nevertheless found the strength to write a well-documented letter defending the program and get it signed by over thirty of Quebec's top specialists in medicine and occupational health. The letter was published within days of the employer attack and given plenty of publicity. Just after Robert died, the government announced it would not reconsider the program, although it was still looking at other aspects of the occupational health and safety law. The right to precautionary leave was protected (at least temporarily). We all miss Robert, but I think this victory is a good memorial to him.

Science has a key role to play in bringing employers and the public across the empathy gap. Ergonomists can document the real work process and what goes wrong with it. Industrial hygienists can detect unhealthy environments. And occupational health specialists and epidemiologists can demonstrate the health effects that arise from work. Not only can these scientists find out about occupational health, their credibility can back up workers' requests for changes in their working conditions and their claims for compensation for health damage. And in fact, this often happens. Many occupational health scientists got into this business to support prevention and compensation efforts. Most of them encounter the same barriers I confronted when I was a young researcher trying to prevent radiation exposures, and few of us have any training or contacts that would help us get closer to those community groups that need our expertise. I hope that our university-community alliance for health can be multiplied and that economic support for these programs can be restored.

While I was wondering how to end this book, I met with one of my students who had just finished her Ph.D. thesis. While waiting for her committee to read it, Ève Laperrière was helping her husband out with his ergonomics consulting firm. I knew Ève as a careful, competent scientist and wasn't surprised that she did well at writing reports. But, because she is young, warm, and gentle, I was surprised and impressed with her ability to testify against severe opposition in hotly contested court cases. She described a case she had won, saying, "We save lives!" The claimant, "M. Rousseau," had a joint problem and couldn't do his previous job driving heavy equipment. The employer wanted to fire him for refusing to do his old job, rather than offering him another job. The employer had to show in court that M. Rousseau was able to do the job and had refused, while M. Rousseau maintained that his injury prevented him from driving. Ève had observed the job and detailed all the movements drivers had to make. She couldn't do that by watching M. Rousseau do the job himself – he was unable to do it. So she had to observe another worker, enabling the employer to respond by saying her observations of another worker weren't relevant. In short, a typical Catch-22 experience with the workers' compensation system.

M. Rousseau, broke and in pain, was pitted against a large, aggressive firm of lawyers and depended on Ève to show he was telling the truth. In fact, she told me, he had wept in gratitude when they first met, saying "You're the first person who listened to me." What really irked Ève was that the other side had called in an ergonomist to testify against her and that "Josiane" had said Ève's report was inaccurate because she hadn't observed M. Rousseau at his job. Ergonomists in Quebec are a small community, and she knew Josiane well. Ève was outraged when Josiane testified that the report was full of errors. Hadn't they taken the same courses, and been supervised together by Nicole during their practicum? We discussed whether Josiane (and others who behaved like her) really believed what they were saying, or whether they would just say whatever would earn their fee. Ève concluded by saying that she wanted to go on doing research, so she could bring ever more proof to Josiane.

But I don't think Josiane needs more proof or more research. Josiane might very well use the information to make the workers' jobs worse, not better. Some ergonomists have complained that when they explain to employers how to make changes at the worksite to diminish the risk of musculoskeletal problems, the employers sometimes adopt the changes and then speed up the assembly line, getting the same injury rate but better production.

I think Josiane and the judge are unable to listen to Ève and M. Rousseau for the same reason I was unable to listen to Micheline and Ana María, and M. Lejeune was unable to listen to me. I think part of the problem is that it is not in their interest to listen. Although employers and managers might get some useful hints from their workers about how to organize their jobs, they don't get the big bucks for making work fun, agreeable, or safe. So why would they want to listen to workers?

On the other hand, the public is paying for health care and replacement income for injured workers whose injuries were not prevented or compensated in the current system. When asbestos-exposed workers die of protracted lung diseases, they and their families incur enormous financial and emotional costs. If the families' claims for compensation are refused because judges haven't heard the best science,[8] the whole society picks up the tab. We are all affected when the hospital workers who take care of us and the teachers who educate our children are exhausted and resentful. That store clerk who said her employer didn't care if she smiled at work may have been right, but I was probably not the only customer who was affected by how unhappy she looked. Listening to her would probably have made her healthier as well as giving a lift to the hundreds of people going through her checkout station. I think the invisible barriers to doing worker-friendly science affect most people in society, while protecting a very few.

Right now our governments, influenced by "marketplace" arguments, are making it harder and harder for scientists to make the kinds of choices I was able to make during my career. In fact, at a recent meeting of the university outreach service held with professors in environmental sciences, a newly-hired professor told me: "I'd love to do what you did. But it's not like that any more. We wouldn't get grants and we would lose our jobs." Sad but very likely true. It is even getting harder to get money for basic research.

Yet science itself has a lot to gain from community partnership. I have described how our questions about prolonged standing arose from listening to supermarket checkout clerks and bank tellers, leading to original research questions and eventually to peer-reviewed publications and a new province-wide study. Similarly, contacts with poultry slaughterhouse workers inspired my colleagues Donna Mergler and Nicole Vézina to produce the first reports on the effects of cold temperatures on perimenstrual pain. I can only think that if the biomechanics scientists who model lifting using cadavers and computers spent more time with factory workers, they would realize that breast size affects load distribution and put it in their models. If industrial hygienists hung out with cleaners, they would probably think of more ways to reduce chemical exposures.

Science has a key role to play in bringing employers and the public across the empathy gap. Ergonomists can document the real work process and what goes wrong with it. Industrial hygienists can detect unhealthy environments. Occupational health specialists and epidemiologists can demonstrate the health effects that arise from work. Not only can these scientists find out about occupational health, their credibility can back up workers' requests for changes in their working conditions and their claims for compensation for health damage. And in fact, this often happens. But we would do a better job if we had more information on workers' perspectives. We can only document what we can observe.

Unfortunately, most scientists, even those in occupational health, have little opportunity to spend time with workers. When they do, they encounter the same barriers I confronted when I was a young researcher trying to prevent radiation exposures. They face personal attacks, challenges, and insults. Too often, they have paid a high price in loss of face, research support, and even jobs. A good number of researchers have just barrelled on regardless, using their courage, ingenuity, and persistence to fight in scientific and public venues to make workers' needs more visible. Most would benefit from support systems and networks that would help us get closer to those community groups that need scientific expertise.

Support for "empathetic science" should come from three sources: the academy, workers, and the public. First, researchers need more university-community partnership arrangements like the one my colleagues and I benefit from (grant programs, university chairs, release time). We need programs that guide scientists in their first contacts with community groups and experienced counsellors to help with the initial, often-difficult attempts by each side to understand the other's needs. We need within-university, departmental, and faculty support for the risky adventures involved in developing surprising projects, recognizing that the payoff can be substantial in terms of innovative, exciting research ideas. We need our universities to lobby for us so that our funders will develop specific programs to develop community-based research. We need journals that will appropriately evaluate and publish the research results.[9] We need research networks that put us in touch with community-based scientists.[10]

Second, we need specific funding for scientist-community partnerships, on a long-term, program basis, where the evaluation of research proposals includes not only peer review but also community-based critiques of the proposal and the research interpretations. That is, scientists need time and frameworks so that relationships can develop on a long-term basis.

Third, we need real support from our community partners. We most of all need them to be sceptical when we are criticized by our peers, and to have the reflex of looking at the evidence. I recently had the unpleasant experience of hearing from several union people that the research on which our defence of precautionary leave was based was not published in good journals.[11] Excuse me? The *American Journal of Public Health*, *American Journal of Epidemiology*, not good journals? And wherever had they heard this? From the employers' organization that was trying to do away with precautionary leave. This was not the only time that I heard union health and safety experts adopt a hypercritical attitude toward their "own" scientists. We need our community partners to educate themselves about the misuse of science and pseudoscientific arguments so they can join us in lobbying for more, better, and more useful science.

We also need the community organizations to ask governments for specific funding programs for community-oriented projects. Right now, there is lobbying and, occasionally, funding for very specific community-oriented calls – for research on why people gamble, on specific diseases, on how to reduce sodium in the diet. But there is no more funding for community-oriented research *programs*, sustained efforts to contribute to understanding thoroughly the needs and environments of homeless people or people exposed to manganese pollution or pregnant miners.

I hope people will wake up and realize that it is in their interest to create more ways to favour community-based research. I hope that scientists will come to understand that community-based science yields important information that is not available through other types of research. I hope that workers will recognize that they have a right to insist on respect for their knowledge and their efforts.

Notes

Preface

1 University employee Michel Lizée set up the agreement, and his brilliance in conceiving its safeguards has preserved it in the face of frequent attacks and allowed it to evolve. As of this writing, it has lasted thirty-eight years and has enlarged to include another union confederation. A similar agreement was set up with community women's groups four years later and is also still active. For an explanation of the union agreement, see Donna Mergler, "Worker Participation in Occupational Health Research: Theory and Practice," *International Journal of Health Services* 17 (1987): 151–67. For more recent projects and an overview of the outreach activities, see www.actualites.uqam.ca/2014/4306-sac-linterface-entre-chercheurs-groupes-sociaux.

2 Centre de recherche interdisciplinaire sur la biologie, la santé, la société et l'environnement, now directed by Johanne Saint-Charles.

3 The now defunct Conseil québécois de recherche sociale.

4 Karen Messing, ed., *Integrating Gender in Ergonomic Analysis* (Brussels: European Trade Union Institute, 1999); Karen Messing and Piroska Ostlin, *Gender Equality, Work and Health: A Review of the Evidence* (Geneva: World Health Organization, 2006). www.who.int/gender/documents/Genderworkhealth.pdf.

Chapter 1 Factory Workers

1 Since, to my mind, the responses of these academics result from systemic rather than individual factors, I have used pseudonyms to identify them. The systemic factors will be discussed throughout the book.

2 The chromosome was tricentric, which is pretty rare and can be caused by radiation. We referred Professor Ivy to the worker for permission, but we have never seen the book with the picture of his chromosome.

3 Louis-Gilles Francoeur, "Scorries radioactives sur la rive sud," *Le Devoir*, May 5, 1993, A1.

4 This contract provision was later to become law in Quebec, where pregnant or nursing workers exposed to a danger for themselves or their fetus or child can apply for reassignment to tasks with no such dangers. Failing such reassignment, they are entitled to paid leave.

5 At that time, film badges were used to measure radiation exposure. The film darkened when exposed to radiation and specialized laboratories could calculate the amount of exposure and report it to the employer.

6 This doesn't necessarily mean that the readings were inaccurate in general. A worker in the laboratory that processed the dosimeters explained to me that they were instructed to disre-

gard aberrant readings. If one worker's film was entirely black but no one else's was, probably that reading would have been disregarded.

7 Sunlight is a very different kind of radiation with no ability to penetrate the human body. That is why X-rays are used to photograph the inside of the body.

8 See chapter 10 for a discussion of objectivity.

9 Stephanie Premji, Katherine Lippel, and Karen Messing, "On travaille à la seconde! Rémunération à la pièce et santé et sécurité du travail dans une perspective qui tient compte de l'ethnicité et du genre," *PISTES* 10, no. 1 (2008). www.pistes.uqam.ca/v10n1/articles/v10n1a2.htm.

10 David Michaels, *Doubt is Their Product* (Don Mills: Oxford University Press, 2008); Naomi Conway Oreskes, *Merchants of Doubt* (New York: Bloomsbury Press, 2010), Chapter 5.

Chapter 2 The Invisible World of Cleaning

Epigraph: "Ils ne verront pas le bon dieu de leur sainte vie, d'avoir fait ça noir."

1 We were lucky to have Abby to coach us. It has been amazing to me that this step is so often omitted when scientists develop questionnaires.

2 Daniel Tierney, Patrizia Romito, and Karen Messing, "She Ate Not the Bread of Idleness: Exhaustion is Related to Domestic and Salaried Work of Hospital Workers in Quebec," *Women and Health* 16 (1990): 21–42.

3 Agathe Croteau, Sylvie Marcoux, and Chantal Brisson, "Work Activity in Pregnancy, Preventive Measures, and the Risk of Delivering a Small-for-Gestational-Age Infant," *American Journal of Public Health* 96, no. 5 (2006): 846–55.

4 Ana María Seifert, Karen Messing, and Diane Elabidi, "Analyse des communications et du travail des préposées à l'accueil d'un hôpital pendant la restructuration des services," *Recherches féministes* 12, no. 2 (1999): 85–108.

5 An Act Respecting Occupational Health and Safety, R.S.Q., ch. S-2.1 (1979). www2.publicationsduquebec.gouv.qc.ca.

6 Melissa A. McDiarmid and Patricia Gucer, "The GRAS Status of Women's Work," *Journal of Occupational and Environmental Medicine* 43, no. 8 (2001): 665–69.

7 Karen Messing and W.E.C. Bradley, "In Vivo Mutant Frequency Rises Among Breast Cancer Patients After Exposure to High Doses of Gamma-Radiation," *Mutation Research* 152, no. 1 (1985): 107–12.

8 It had previously been organized into "cleaning – men" and "cleaning – women" but advertising or designating jobs according to gender became illegal in the 1960s.

9 Karen Messing, Céline Chatigny, and Julie Courville, "'Light' and 'Heavy' Work in the Housekeeping Service of a Hospital," *Applied Ergonomics* 29, no. 6 (1998): 451–59.

10 Karen Messing, "Do Men and Women Have Different Jobs Because of Their Biological Differences?" *International Journal of Health Services* 12 (1982): 43–52.

11 Scientists generally use "sex" to talk about biological characteristics and "gender" to refer to social roles. Despite the impossibility of clearly dividing real-world phenomena into social versus biological, it is convenient to use this terminology. But see Anne Fausto-Sterling, "The Bare Bones of Sex: Part 1 – Sex and Gender," *Signs* (Winter 2005): 1491–1527.

12 *On est tanné d'être traité comme des taureaux!*

13 Karen Messing, Ana María Seifert, Jocelyne Ferraris, Joel Swarz, and W.E.C. Bradley, "Mutant Frequency of Radiotherapy Technicians Appears to Reflect Recent Dose of Ionizing Radiation," *Health Physics* 57 (1989): 537–44.

14 See Karen Messing and Donna Mergler, "Determinants of Success in Obtaining Grants for Action-Oriented Research in Occupational Health," *Proceedings of the American Public Health Association held in Las Vegas, Nevada, September 28–October 2, 1986*, 91. Thanks to Karla Pearce of APHA for her help in retrieving this abstract.

15 Possibly because of negative effects of tobacco on cell growth.

16 Gina B. Kolata, "Testing for Cancer Risk," *Science* 207, no. 4434 (1980): 967–69.

17 Susan M. Reverby, "Invoking Tuskegee: Problems in Health Disparities, Genetic Assumptions, and History," *Journal of Health Care of the Poor and Underserved* 21, no. 3 Suppl. (2010): 26–34.

18 And our own work later showed that the test would probably not work to identify individuals who were likely to get cancer. While the number of mutant cells found in a radiation-exposed group was related to exposure, the number of mutant cells found in any individual was a result of an interaction between the exposure and the state of the cell that happened to be hit at the time. See for example W.E.C. Bradley and Karen Messing, "Fluctuations in Mutant Frequency in CHO Cells Exposed to Very Low Doses of Ionizing Radiation are Due to Selection of Radioresistant Subpopulations," *Carcinogenesis* 7 (1986): 1451–55.

19 Ana María Seifert, Christian Demers, Hélène Dubeau, and Karen Messing, "HPRT-Mutant Frequency and Lymphocyte Characteristics of Workers Exposed to Ionizing Radiation on a Sporadic Basis: A Comparison of Two Exposure Indicators, Job Title and Dose," *Mutation Research* 319, no. 1 (1993): 61–70.

20 Karen Messing and W.E.C. Bradley, "In Vivo Mutant Frequency Rises Among Breast Cancer Patients After Exposure to High Doses of Gamma-Radiation," *Mutation Research* 152, no. 1 (1985): 107–12.

21 The test we were using is still being applied by concerned scientists to detection and prevention of environmentally induced mutations in humans. See M.A. McDiarmid, S.M. Engelhardt, M. Oliver, P. Gucer, P.D. Wilson, R. Kane, A. Cernich, B. Kaup, L. Anderson, D. Hoover, L. Brown, R. Albertini, R. Gudi, D. Jacobson-Kram, and K.S. Squibb, "Health Surveillance of Gulf War I Veterans Exposed to Depleted Uranium: Updating the Cohort," *Health Physics* 93, no. 1 (2007): 60–73.

22 A description of the approach to ergonomics that Nicole and, eventually, I learned can be found in Karen Messing, Ana María Seifert, Nicole Vézina, Ellen Balka, and Céline Chatigny, "Qualitative Research Using Numbers: Analysis Developed in France and Used to Transform Work in North America," *New Solutions: A Journal of Environmental and Occupational Health Policy* 15, no. 3 (2005): 245–260.

23 Christian Demers, Nicole Vézina, and Karen Messing, "Le travail en présence de radiations ionisantes dans des laboratoires universitaires," *Radioprotection* 26 (1991): 387–95.

24 Dejours founded the study of work psychodynamics: how workers use collective strategies to protect themselves from risk at work or from fear of risk at work. For example, Dejours studied the ways nuclear power plant workers played dangerous games at their plants in order to deal with their fear of exposure and of making mistakes. See Christophe Dejours, *Travail-Usure mentale,* 2nd ed. (Paris: Bayard, 1993).

25 Catherine Teiger, Antoine Laville, Jeanne Boutin, Lucien Etxezaharreta, Leonardo Pinsky, Norbert See, and Jacques Theureau, *Les rotativistes: Changer les conditions de travail* (Paris: Éditions de l'ANACT, 1982).

26 Danièle Kergoat, *Les ouvrières* (Paris: Sycomore, 1982).

27 The men told me they had actually held a one-day strike when one of them was assigned to cleaning toilets.

28 Muriel Dimen, "Servants and Sentries – Women, Power, and Social Reproduction in Krióvrisi," *Journal of Modern Greek Studies* 1, no. 1 (1983): 225–42.

29 Karen Messing, Ghislaine Doniol-Shaw, and Chantal Haëntjens, "Sugar and Spice: Health

Effects of the Sexual Division of Labour Among Train Cleaners," *International Journal of Health Services* 23, no. 1 (1993): 133–46.

30 Karen Messing, Chantal Haëntjens, and Ghislaine Doniol-Shaw, "L'invisible nécessaire: l'activité de nettoyage des toilettes sur les trains de voyageurs en gare," *Le travail humain* 55 (1992): 353–70.

31 Prix Jacques Rousseau from the Association francophone pour le savoir.

32 Lucie Dagenais was then in charge of the health and safety sessions and did a wonderful job of developing ways to access the latest scientific knowledge for workers.

33 Karen Messing, Céline Chatigny, and Julie Courville, "'Light' and 'Heavy' Work in the Housekeeping Service of a Hospital," *Applied Ergonomics* 29, no. 6 (1998): 451–59.

34 The suggestion was probably accepted so readily in part because pay equity legislation had increased pay for the women's jobs.

35 Bénédicte Calvet, Vanessa Couture, Jessica Riel, and Karen Messing, "Work Organization and Gender Among Hospital Cleaners in Quebec After the Merger of 'Light' and 'Heavy' Work Classifications," *Ergonomics* 55, no. 2 (2012): 160–72.

36 Élise Fortin, Charles Frenette, Suzanne Gingras, Marie Gourdeau, and Bruno Hubert, *Surveillance des diarrhées associées à Clostridium difficile au Québec* (Québec: Institut national de santé publique du Québec, 2005). www.inspq.qc.ca; Presentation by Alain Poirier, Director of the Institut national de santé publique du Québec, November 2006. www.msss.gouv.qc.ca/sujets/prob_sante/nosocomiales/index.php?presentation.

37 There were three architects, two each of nurses, engineers, microbiologists, and public health scientists, and one "administrative procedures analyst." See Ministère de la santé et des services sociaux du Québec, *Prévention et contrôle des infections nosocomiales. Principes généraux d'aménagement* (Québec: Ministère de la santé et des services sociaux, 2009). http://publications.msss.gouv.qc.ca.

Chapter 3 Standing Still

Epigraph: "Il va pas me donner un massage dans le dos, donc je ne me plains pas."

1 Nicole Vézina, Céline Chatigny, and Karen Messing, "A Manual Materials Handling Job: Symptoms and Working Conditions Among Supermarket Cashiers," *Chronic Diseases in Canada/Maladies chroniques au Canada* 15 (1994): 17–22.

2 Karen Messing and Sophie Boutin, "La reconnaissance des conditions difficiles dans les emplois des femmes et les instances gouvernementales en santé et en sécurité du travail," *Relations industrielles/Industrial Relations* 52 (1997): 333–62.

3 France Tissot, Karen Messing, and Susan Stock, "Standing, Sitting and Associated Working Conditions in the Quebec Population in 1998," *Ergonomics* 48 (2005): 249–69.

4 In fact, prolonged sitting is not good either. What's best is when workers can control how much sitting and standing they do.

5 Ronald C. Plotnikoff and Nandini Karunamuni, "Reducing Sitting Time: The New Workplace Health Priority" (Editorial), *Archives of Environmental and Occupational Health* 67 (2012): 125–27.

6 David Baty, Peter W. Buckle, and David A. Stubbs, "Posture Recording by Direct Observation, Questionnaire Assessment and Instrumentation: A Comparison Based on a Recent Field Study." In *The Ergonomics of Working Postures: Models, Methods and Cases*, ed. Nigel Corlett, John R. Wilson, and Ilija Manenica (London: Taylor & Francis, 1986), 283–92; Christina Wiktorin, Lena Karlqvist, and Jorgen Winkel, "Validity of Self-Reported Expo-

sures to Work Postures and Manual Materials Handling. Stockholm MUSIC I Study Group," *Scandinavian Journal of Work Environment & Health* 19 (1993): 208–14.

7 Sven Erik Mathiassen, "Diversity and Variation in Biomechanical Exposure: What Is It, and Why Would We Like to Know?" *Applied Ergonomics* 37 (2006): 419–27.

8 Karen Messing, Laura Punnett, and Eira Viikari-Juntura, "Åsa Kilbom, 1938–2005," *Applied Ergonomics* 37 (2006): 681–82.

9 Åsa Kilbom, "Physical Training With Submaximal Intensities in Women. III. Effect on Adaptation to Professional Work," *Scandinavian Journal of Clinical and Laboratory Investigation* 28 (1971): 331–43.

10 Karen Messing and Åsa Kilbom, "Standing and Very Slow Walking: Foot Pain-Pressure Threshold, Subjective Pain Experience and Work Activity," *Applied Ergonomics* 32 (2001): 81–90.

11 Ève Laperrière, Suzy Ngomo, Marie-Christine Thibault, and Karen Messing, "Indicators for Choosing an Optimal Mix of Major Working Postures," *Applied Ergonomics* 37 (2006): 349–57; Suzy Ngomo, Karen Messing, Hélène Perreault, and Alain-Steve Comtois, "Orthostatic Symptoms, Blood Pressure and Working Postures of Factory and Service Workers Over an Observed Work Day," *Applied Ergonomics* 39 (2008): 729–36; Marie Laberge and Nicole Vézina, "Un banc assis-debout pour les caissières: une solution pour réduire les contraintes de la position debout?" *Travail et santé* 14, no. 2 (1998): 42–48.

12 Karen Messing, "La place des femmes dans les priorités de recherche en santé au travail au Québec," *Relations industrielles/Industrial Relations* 57 (2002): 660–86.

13 Karen Messing, Katherine Lippel, Ève Laperrière, and Marie-Christine Thibault, "Pain Associated with Prolonged Constrained Standing: The Invisible Epidemic." In *Occupational Health and Safety: International Influences and the "New" Epidemics*, ed. Chris L. Peterson and Claire Mayhew (Baywood: Amityville, New York, 2005), 139–57.

14 Niklas Krause, John W. Lynch, George A. Kaplan, Richard D. Cohen, Riitta Salonen, and Jukka T. Salonen, "Standing at Work and Progression of Carotid Atherosclerosis," *Scandinavian Journal of Work, Environment and Health* 26 (2000): 227–36; Finn Tüchsen, Harald Hannerz, Hermann Burr, and Niklas Krause, "Prolonged Standing at Work and Hospitalisation due to Varicose Veins: A 12 Year Prospective Study of the Danish Population," *Occupational and Environmental Medicine* 62 (2005): 847–50.

15 France Tissot, Karen Messing, and Susan Stock, "Studying Relations Between Low Back Pain and Working Postures Among Those Who Stand and Those Who Sit Most of the Work Day," *Ergonomics* 52 (2009): 1402–18.

16 When I asked the man about shoes, he replied out that he didn't earn enough money to buy good shoes and that this was a problem for him since his shoes had to be replaced every couple of months because of all the hard use they got.

17 Anna Sam, *Checkout Girl: A Life Behind the Register* (New York: Sterling, 2008).

18 *Tu vois, si tu ne travailles pas bien à l'école, tu deviendras caissière comme la dame.* Sam, *Checkout Girl*, 108.

19 France Tissot, Karen Messing, and Susan Stock, "Standing, Sitting and Associated Working Conditions."

20 Karen Messing, *One-Eyed Science: Occupational Health and Women Workers* (Philadelphia: Temple University Press, 1998).

21 Vanessa Couture, "Adaptations cardiovasculaires et inconfort lors du maintien d'une posture debout prolongée," (master's thesis, Université du Québec à Montréal, 2008).

22 Karen Messing, Sylvie Fortin, Geneviève Rail, and Maude Randoin, "Standing Still: Why North American Workers are Not Insisting on Seats Despite Known Health Benefits," *International Journal of Health Services* 35 (2005): 745–63.

23 Karen Messing, France Tissot, Vanessa Couture, and Stephanie Bernstein, "Strategies for Work/Life Balance of Women and Men with Variable and Unpredictable Work Hours in the Retail Sales Sector in Québec, Canada," *New Solutions: A Journal of Environmental and Occupational Health Policy* (in press).

Chapter 4 The Brains of Low-Paid Workers

Epigraph: "Pour être serveuse, tu as besoin de toute ta tête."

1 Catherine Teiger and Antoine Laville, *Les rotativistes: Changer les conditions de travail* (Montrouge, France: Agence pour l'Amélioration des Conditions de Travail, 1982).
2 Nicole Vézina, Daniel Tierney, and Karen Messing, "When is Light Work Heavy? Components of the Physical Workload of Sewing Machine Operators Which May Lead to Health Problems," *Applied Ergonomics* 23 (1992): 268–76.
3 F. Guérin, A. Laville, F. Daniellou, J. Durrafourg, and A. Kerguelen, *Understanding and Transforming Work: The Practice of Ergonomics* (Lyon: ANACT Network Editions, 2007).
4 Marie-Claude Plaisintin and Catherine Teiger, "Les contraintes du travail dans les travaux répétitifs de masse et leurs conséquences pour les travailleuses." In *Les effets des conditions de travail sur la santé des travailleuses*, ed. Jeanne-Anne Bouchard (Montréal: Confédération des Syndicats Nationaux, 1984).
5 Along with Ghislaine Doniol-Shaw, Alain Garrigou, and François Daniellou.
6 Catherine Teiger and C. Bernier, "Ergonomic Analysis of Work Activity of Data Entry Clerks in the Computerized Service Sector can Reveal Unrecognized Skills," *Women and Health* 18, no. 3 (1992): 67–77.
7 Nicole Vézina, Susan R. Stock, Yves Saint-Jacques, Micheline Boucher, Jacques Lemaire, and Conrad Trudel, *Problèmes musculo-squelettiques et organisation modulaire du travail dans une usine de fabrication de bottes ou "Travailler en groupe, c'est de l'ouvrage"* (Montréal: Direction de la santé publique, Régie Régionale de la Santé et des Services Sociaux de Montréal-Centre, 1998).
8 Chantal Brisson, Alain Vinet, Michel Vézina, and Suzanne Gingras, "Effect of Duration of Employment in Piecework on Severe Disability among Female Garment Workers," *Scandinavian Journal of Work, Environment and Health* 15 (1989): 329–34.
9 Some of the story of the boot factory is told in the film *Asking Different Questions*, directed by Gwynne Basen and Erna Buffie (Canada: National Film Board of Canada, 1996).
10 Patrick G. Dempsey and Alfred J. Filiaggi, "Cross-Sectional Investigation of Task Demands and Musculoskeletal Discomfort among Restaurant Wait Staff," *Ergonomics* 49 (2006): 93–106; Chyuan Jong-Yu, Chung-Li Du, Wen-Yu Yeh, and Chung-Yi Li, "Musculoskeletal Disorders in Hotel Restaurant Workers," *Occupational Medicine* 54 (2004): 55–57; Hermann Hannerz, Finn Tüchsen, and Tage S. Kristensen, "Hospitalizations among Employees in the Danish Hotel and Restaurant Industry," *European Journal of Public Health* 12 (2002): 192–97.
11 Ève Laperrière, Karen Messing, and Rennée Bourbonnais, "Pour être serveuse, tu as besoin de toute ta tête: Efforts et reconnaissance dans le service de table au Québec," *Travailler* 23 (2010): 27–58.
12 Michael Lynn, "Determinants and Consequences of Female Attractiveness and Sexiness: Realistic Tests with Restaurant Waitresses," *Archives of Sex Behavior* 38 (2009): 737–45.
13 Ève Laperrière, "Étude du travail de serveuses de restaurant" (Ph.D. diss., Université du Québec à Montréal, 2014).

14 I searched the Pub Med database for "waiter OR waitress OR food server NOT tobacco" in the title or abstract of an article for the years 2000–2010, with the search limited to human adults. I searched for "restaurant work" and "restaurant occupation" with the same limitations. I also searched the database for (food service OR waiter OR restaurant OR waitress) AND occupation AND health for the years 1995–2010, with the same limitations.

15 A population-based study is one that does not select a specific group at risk, such as wait staff, but looks at a more general group, such as patients hospitalized for cancer, to see whether wait staff appear more often than they should, compared to other occupations.

16 Béatrice Cahour and Barbara Pentimalli, "Conscience périphérique et travail coopératif dans un café-restaurant," *Activités* 2 (2005): 50–75. www.activites.org/v2n1/cahour.pdf; Gro Ellen Mathisen, Ståle Einarsen, and Reidar Mykletun, "The Occurrences and Correlates of Bullying and Harassment in the Restaurant Sector," *Scandinavian Journal of Psychology* 49 (2008): 59–68.

17 Valerie K. York, Laura A. Brannon, Carol W. Shanklin, Kevin R. Roberts, Amber D. Howells, and Betsy B. Barrett, "Foodservice Employees Benefit from Interventions Targeting Barriers to Food Safety," *Journal of the American Dietetic Association* 109 (2009): 1576–81.

Chapter 5 Invisible Teamwork

1 Ana María Seifert and Karen Messing, "Cleaning Up After Globalization: An Ergonomic Analysis of Work Activity of Hotel Cleaners," *Antipode* 38 (2006): 557–77.

2 Fédération des travailleurs et travailleuses du Québec (Québec Federation of Labour).

3 Centrale de l'enseignement du Québec, now the CSQ (Centrale des syndicats du Québec).

4 Karen Messing, ed., *Integrating Gender in Ergonomic Analysis* (Brussels: European Trade Union Institute, 1999).

5 Karen Messing and Piroska Ostlin, *Gender Equality, Work and Health: A Review of the Evidence* (Geneva: World Health Organization, 2006). www.who.int/gender/documents/.

6 Ana María Seifert, Karen Messing, and Lucie Dumais, "Star Wars and Strategic Defense Initiatives: Work Activity and Health Symptoms of Unionized Bank Tellers During Work Reorganization," *International Journal of Health Services* 27 (1997): 455–77.

7 Ana María Seifert, Karen Messing, and Lucie Dumais, "Star Wars and Strategic Defense Initiatives."

8 Cynthia J. Cranford, Leah F. Vosko, and Nancy Zukewich, "The Gender of Precarious Employment in Canada," *Relations Industrielles/Industrial Relations* 58 (2003): 454–82.

9 Karen Messing and Diane Elabidi, "Desegregation and Occupational Health: How Male and Female Hospital Attendants Collaborate on Work Tasks Requiring Physical Effort," *Policy and Practice in Health and Safety* 1 (2003): 83–103.

10 Ana María Seifert and Karen Messing, "Looking and Listening in a Technical World: Effects of Discontinuity in Work Schedules on Nurses' Work Activity," *PISTES* 6 (2004). www.pistes.uqam.ca/v6n1/articles/v6n1a3en.htm.

11 Apparently since 2012 in Quebec rehospitalization figures are published that include all causes.

12 Céline Chatigny and Nicole Vézina, "Analyse du travail et apprentissage d'une tâche complexe; étude de l'affilage du couteau dans un abattoir," *Le Travail Humain* 58 (1995): 229–52.

13 Nicole Vézina, Johanne Prévost, and Alain Lajoie, *Coupera ou coupera pas?* (Montréal: Université du Québec à Montréal. Service de l'audiovisuel, 1997). Videocassette (VHS).

Chapter 6 Home Invasion

1 "La FTQ dénonce « le droit à l'escalope » à toute heure du jour ou de la nuit," *Le Devoir*, December 7, 2006, A2.

2 Imelda S. Wong, Christopher B. McLeod, and Paul A. Demer, "Shift Work Trends and Risk of Work Injury among Canadian Workers," *Scandinavian Journal of Work, Environment and Health* 36 (2011): 54–61.

3 International Agency for Research on Cancer, *IARC Monographs on the Evaluation of Carcinogenic Risks to Humans: Painting, Firefighting, and Shiftwork* (Lyon: International Agency for Research on Cancer, 2010); Agathe Croteau, Sylvie Marcoux, and Chantal Brisson, "Work Activity in Pregnancy, Preventive Measures, and the Risk of Delivering a Small-for-Gestational-Age Infant," *American Journal of Public Health* 96, no. 5 (2006): 846–55.

4 Alwin Van Drongelen, Cécile R.L. Boot, Suzanne L. Merkus, Tjabe Smid, and Allard J. van der Beek, "The Effects of Shift Work on Body Weight Change: A Systematic Review of Longitudinal Studies," *Scandinavian Journal of Work, Environment and Health* 37 (2011): 263–75.

5 Linda Duxbury and Chris Higgins, *Work–Life Conflict in Canada in the New Millennium:Key Findings and Recommendations From The 2001 National Work–Life Conflict Study* (Ottawa: Health Canada, 2009). www.hc-sc.gc.ca.

6 Johanne Prévost and Karen Messing, "Stratégies de conciliation d'un horaire de travail variable avec des responsabilités familiales," *Le travail humain* 64 (2001): 119–43.

7 Béatrice Barthe, Linda Abbas, and Karen Messing, "Strategies Used by Women Workers to Reconcile Family Responsibilities with Atypical Work Schedules in the Service Sector," *Work* 40 (Supplement, 2012): S47–S58.

8 We were asked not to identify the type of store.

9 Catherine Des Rivières and Isabelle Courcy, *"Work-Family Balance": Portrait of Recent Articles and Proposals for Future Research* (Montréal: Université du Québec à Montréal, Department of Sociology, 2010).

10 "Je ne vois juste pas à quoi ça donne de remplir ce formulaire quand je sais très bien que l'UQAM en a rien à foutre que je crois que je n'ai pas assez d'heures dans mon département de [department] au [store name] [store location] Québec."

Chapter 7 Teachers and Numbers

1 *Business Dictionary*, "What Gets Measured Gets Improved." www.businessdictionary.com.

2 Jessica Riel and Karen Messing, "Counting the Minutes: Balancing Work and Family Among Secondary School Teachers in Québec," *Work* 40 (Supplement, 2011): S59–S70.

3 "François Legault et l'éducation: trois idées contestées," *La Presse*, April 13, 2011. www.cyberpresse.ca.

4 Pasi Sahlberg, *Finnish Lessons* (New York: Teachers College Press/Columbia University, 2011).

5 Helen F. Ladd, "Teacher Labor Markets in Developed Countries," *Future Child* 17 (2007): 201–17.

6 Karen Messing, Ana María Seifert, and Evelin Escalona, "The 120-Second Minute: Using Analysis of Work Activity to Prevent Psychological Distress Among Elementary School Teachers," *Journal of Occupational Health Psychology* 2 (1997): 45–62.

7 Karen Messing, Ana María Seifert, and Evelin Escalona, "The 120-Second Minute."

8 Jacqueline Dionne-Proulx, "Le stress au travail et ses conséquences potentielles à long terme: le cas des enseignants québécois," *Revue canadienne de l'éducation* 20 (1995): 146–55; Nathalie Houlfort and Frédéric Sauvé, *Santé psychologique des enseignants de la Fédération autonome de l'enseignement* (Montréal: École nationale d'administration publique, 2010).
9 Karen Messing, Ana María Seifert, and Evelin Escalona, "The 120-Second Minute."
10 Lysiane Gagnon, "Pas le gout de travailler [Don't feel like working]," *La Presse,* May 7, 1994, B3.
11 Agnès Gruda, "20 journees pedagogiques sur 200 [20 pedagogical days out of 200]," *La Presse,* June 9, 1994, B2.
12 Jessica Riel and Karen Messing, "Counting the Minutes."
13 Karen Messing and Ana María Seifert. "'On est là toutes seules': Contraintes et stratégies des femmes en contrat à durée déterminée dans l'enseignement aux adultes," *Travailler* 7 (2002): 147–66.
14 J. Mukamurera, "L'insertion professionnelle chez les jeunes : un problème complexe qui commande une stratégie globale." In *Actes de colloque – Pour une insertion réussie dans la profession enseignante : passons à l'action!* (Québec: Ministère de l'Éducation du Québec, COFPE, CRIFPE, 2004); M. Perreault, M., "Un salarié de la CSQ sur trois est en situation à risque d'épuisement professionnel au travail." Union press release, August 24, 2004. www.csq.qc.net. Consulted July 21, 2010.
15 Esther Cloutier, Élise Ledoux, Madeleine Bourdouxhe, Hélène David, Isabelle Gagnon, and François Ouellet, "Restructuring of the Québec Health Network and its Effects on the Profession of Home Health Aides and their Occupational Health and Safety," *New Solutions: A Journal of Environmental and Occupational Health Policy* 17 (2007): 83–95.
16 Karen Messing, Ana María Seifert, Nicole Vézina, Ellen Balka, and Céline Chatigny, "Qualitative Research Using Numbers: Analysis Developed in France and Used to Transform Work in North America," *New Solutions: A Journal of Environmental and Occupational Health Policy* 15 (2005): 245–60.
17 Ève Laperrière, Suzy Ngomo, Marie-Christine Thibault, and Karen Messing, "Indicators for Choosing an Optimal Mix of Major Working Postures," *Applied Ergonomics* 37 (2006): 349–57.

Chapter 8 Becoming a Scientist

1 Jonathan Beckwith, *Making Genes, Making Waves: A Social Activist in Science* (Cambridge, MA: Harvard University Press, 2002).
2 Allen Fenichel and David Mandel, *The Academic Corporation: Justice, Freedom, and the University* (Montreal and New York: Black Rose Books, 1987).
3 Now the Quebec government supplies care at $7/day, although there are not enough spaces for all children.
4 It turns out the Montreal Day Nursery had a tradition of being particularly rigid. See Donna Varga, *Constructing the Child: A History of Canadian Day Care* (Toronto: James Lorimer, 1997).
5 Chromosomes, made of DNA organized into genes, are in the cell nucleus. The structure of the DNA determines the structure of the enzymes made in the rest of the cell, called the cytoplasm. But it is true that reactions taking place in the cytoplasm determine when and how the DNA will be translated into enzymes.
6 Institut de recherche Robert Sauvé en santé et en sécurité du travail.

7 As opposed to remedying the effects of health problems, for example by developing a new type of hearing aid or analyzing data on workplace accidents. One member of the IRSST administration reproached us with, "You people want to prevent problems that haven't even happened yet!"

8 Karen Messing, "La place des femmes dans les priorités de recherche en santé au travail au Québec." *Relations industrielles/Industrial Relations* 57 (2002): 660–86. This decision has recently been reversed.

9 Chris Turner, *The War on Science* (Vancouver: Greystone Books, 2014).

Chapter 9 Crabs, Pain, and Sceptical Scientists

1 Dana Howse, Denise Gautrin, Barbara Neis, André Cartier, Lise Horth-Susin, Michael Jong, and Mark C. Swanson, "Gender and Snow Crab Occupational Asthma in Newfoundland and Labrador, Canada," *Environmental Research* 101 (2006): 163–74; Marie Eve Major and Nicole Vézina, "Ergonomic Study of Seasonal Work and its Impacts on Soft-Tissue Injuries and Strategies of Women Workers in Crab Processing Plants." Communication presented at the annual congress of the Canadian Association for Research in Work and Health, St. John's, Newfoundland, June 10, 2006.

2 Pierre Chrétien, "Quebec Experience in Diagnosis and Management of WMSDs in Crab Plants." Communication presented at the annual congress of the Canadian Association for Research in Work and Health, St. John's, Newfoundland, June 10, 2006.

3 Bradley Evanoff to the OEM-L listserve, May 1997. Quoted with permission.

4 Geert Crombez, Christopher Eccleston, Stefaan Van Damme, Johan W.S. Vlaeyen, and Paul Karoly, "Fear-Avoidance Model of Chronic Pain: The Next Generation," *Clinical Journal of Pain* 28 (12): 475–83.

5 Institut de recherche Robert-Sauvé en santé et en sécurité du travail.

6 Timothy H. Wideman and Michael J.L. Sullivan, "Development of a Cumulative Psychosocial Factor Index for Problematic Recovery Following Work-Related Musculoskeletal Injuries," *Physical* Therapy 92 (2011): 58–68.

7 Michael J.L. Sullivan, Maureen Simmonds, and Ana Velly, *Pain, Depression, Disability and Rehabilitation Outcomes* (Montréal: Institut de recherche Robert-Sauvé en santé et en sécurité du travail, 2011), iii.

8 Timothy H. Wideman and Michael J.L. Sullivan, "Differential Predictors of the Long-Term Levels of Pain Intensity, Work Disability, Healthcare Use, and Medication Use in a Sample of Workers' Compensation Claimants," *Pain* 152 (2011), 376.

9 Michael J. Sullivan, Scott R. Bishop, and Jayne Pivik, "The Pain Catastrophizing Scale: Development and Validation," *Psychological Assessment* 7 (1995): 524–32.

10 Tampa Scale for Kinesiophobia, www.tac.vic.gov.au/upload/tampa_scale_kinesiophobia.pdf.

11 Marie-France Coutu, Raymond Baril, Marie-Josée Durand, Daniel Côté, and Geneviève Cadieux, "Clinician – Patient Agreement about the Work Disability Problem of Patients Having Persistent Pain: Why it Matters," *Journal of Occupational Rehabilitation* 23 (2013): 82–92; Cammie Chaumont Menéndez, Benjamin C. Amick III, Mark Jenkin, Cyrus Caroom, Michelle Robertson, Fred Gerr, J. Steven Moore, Ronald B. Harrist, and Jeffrey N. Katz, "A Validation Study Comparing Two Self-Reported Upper Extremity Symptom Surveys with Clinical Examinations for Upper Extremity Musculoskeletal Disorders," *Work* 43 (2012): 293–302.

12 Enquête Québécoise sur les Conditions de Travail, d'Emploi et de Santé et Sécurité du Travail.

13 In the interest of full disclosure, I should say that I am a co-author of the three chapters that dealt respectively with working conditions, work-family balancing, and musculoskeletal disorders.

14 CNW, "Enquête québécoise sur les conditions de travail, d'emploi et de santé et sécurité du travail – Des conclusions non justifiables, selon les associations patronales du Québec." www.cnw.ca/en.

15 " . . . soulèvent donc de sérieuses questions quant à sa valeur scientifique."

16 $831,000 in direct grants plus the donated labour of twenty-two scientists and twenty reviewers.

17 "l'étude est essentiellement une enquête de perceptions menée uniquement auprès des travailleurs, et non pas une analyse de données objectives et de faits établis."

18 Susan R. Stock, R. Fernandes, Alain Delisle, and Nicole Vézina, "Reproducibility and Validity of Workers' Self-Reports of Physical Work Demands," *Scandinavian Journal of Work, Environment and Health* 31 (2005): 409–37.

19 Ève Laperrière, Vanessa Couture, Susan R. Stock, and Karen Messing, "Validation of Questions on Working Posture Among Standing Workers in Québec," *International Journal of Industrial Ergonomics* 35 (2005): 371–78.

20 Esa-Pekka Takala, Irmeli Pehkonen, Michael Forsman, G.A. Hansson, Svend Erik Mathiassen, W. Patrick Neumann, Gisela Sjøgaard, Kaj BoVeiersted, Rolf H. Westgaard, and Jorgen Winkel, "Systematic Evaluation of Observational Methods Assessing Biomechanical Exposures at Work," *Scandinavian Journal of Work, Environment and Health* 36 (2010): 3–24.

21 John A. Ioannidis, "Why Most Published Research Findings are False," *PLoS Medicine* 2, no. 8 (2005): e124. www.plosmedicine.org.

22 Eve Laperrière, Suzy Ngomo, Marie-Christine Thibault, and Karen Messing, "Indicators for Choosing an Optimal Mix of Major Working Postures," *Applied Ergonomics* 37 (2006): 349–57.

23 Karen Messing and Jean-Pierre Reveret, "Are Women in Female Jobs for their Health? Working Conditions and Health Symptoms in the Fish Processing Industry in Québec," *International Journal of Health Services* 13 (1983): 635–47; Donna Mergler, Carole Brabant, Nicole Vézina, and Karen Messing, "The Weaker Sex? Men in Women's Working Conditions Report Similar Health Symptoms," *Journal of Occupational Medicine* 29 (1987): 417–21.

24 Karen Messing, ed., *Integrating Gender in Ergonomic Analysis* (Brussels: European Trade Union Institute, 1999).

25 For a discussion of such studies, see: Annet H. de Lange, Toon W. Tarsi, Michael A.J. Kompier, Irene L.D. Houtman, and Paulien M. Bongers, "Different Mechanisms to Explain the Reversed Effects of Mental Health on Work Characteristics," *Scandinavian Journal of Work, Environment and Health* 31 (2005): 3–14.

Chapter 10 A Statistician's Toes and the Empathy Gap in Scientific Articles

1 Kee-Seng Chia,"'Significant-itis' – An Obsession with the P-value," *Scandinavian Journal of Work, Environment and Health* 23 (1997): 152–54.

2 Thomas R. Waters, Sherry L. Baron, Laurie A. Piacitelli, Vern P. Anderson, Torsten Skov, Marie Haring-Sweeney, David K. Wall, and Lawrence J. Fine, "Evaluation of the Revised NIOSH Lifting Equation. A Cross-Sectional Epidemiologic Study," *Spine* 24 (1999): 386–94.

3 Katherine Lippel, "L'incertitude des probabilités en droit et médecine," *Revue de droit de l'Université de Sherbrooke* 22 (1992): 445–72.

4 Société canadienne des postes et Corbeil et Grégoire Larivière, [1994] C.A.L.P. 285.

5 Katherine Lippel, Karen Messing, Susan Stock, and Nicole Vézina, "La preuve de la causalité et l'indemnisation des lésions attribuables au travail répétitif: rencontre des sciences de la santé et du droit. [Proof of Causality and Compensation of Repetitive Strain Injuries: An Encounter Between Health Sciences and Law]," *Windsor Yearbook of Access to Justice* 17 (1999): 35–86.

6 Michael Foley, Barbara A. Silverstein, Nayak Polissar, and Blazej Neradilek, "Impact of Implementing the Washington State Ergonomics Rule on Employer Reported Risk Factors and Hazard Reduction Activity," *American Journal of Industrial Medicine* 52 (2009): 1–16.

7 Société canadienne des postes et Corbeil et Grégoire Larivière, [1994] C.A.L.P. 285.

8 Katherine Lippel, "Le droit québécois et les troubles musculo-squelettiques: Règles relatives à l'indemnisation et à la prévention," *PISTES* 11 (2010). www.pistes.uqam.ca/v11n2/articles/v11n2a3.htm.

9 Barbara A. Silverstein, J. Lawrence, and Thomas J. Armstrong, "Hand Wrist Cumulative Trauma Disorders in Industry," *British Journal of Industrial Medicine* 43 (1986): 784 (emphasis added).

10 David Michaels, *Doubt is Their Product.* (New York: Oxford University Press, 2010), Chapter 7.

11 Stephanie Premji, Karen Messing, and Katherine Lippel, "Would a 'One-Handed' Scientist Lack Rigor? How Scientists Discuss the Work-Relatedness of Musculoskeletal Disorders in Formal and Informal Communications," *American Journal of Industrial Medicine* 51 (2008): 173–85.

12 David Egilman, "Suppression Bias at the Journal of Occupational and Environmental Medicine," *International Journal of Occupational and Environmental Health* 11 (2005): 202–204.

13 Florence Tubach, Annette Leclerc, Marie-France Landré, and Françoise Pietri-Taleb, "Risk Factors for Sick Leave Due to Low Back Pain: A Prospective Study," *Journal of Occupational & Environmental Medicine* 44 (2002): 458 (emphasis added).

14 Michele Marcus and Fredric Gerr, "Yes, the "One-Handed" Scientist Lacks Rigor – Why Investigators Should Not Use Causal Language When Interpreting the Results of a Single Study," *American Journal of Industrial Medicine* 51 (2008): 795–96.

15 Katherine Lippel, "Compensation for Musculoskeletal Disorders in Quebec: Systemic Discrimination Against Women Workers?" *International Journal of Health Services* 33 (2003): 253–81.

16 Susan Stock, Nektaria Nicolakakis, Hicham Raiq, Karen Messing, Katherine Lippel, and Alice Turcot, "Do Workers Underreport Work Absences for Non-Traumatic Work-Related Musculoskeletal Disorders to Workers' Compensation? Results of a 2007–2008 Survey of the Québec Working Population," *American Journal of Public Health* 104 (2014): e94–e101.

17 Katherine Lippel, Marie-Claire Lefebvre, Chantal Schmidt, and Joseph Caron, *Traiter la réclamation ou traiter la personne ? Les effets du processus sur la santé des personnes victimes de lésions professionnelles* (Montréal: Service aux collectivités de l'Université du Québec à Montréal, 2007).

Chapter 11 Can Scientists Care?

1 Don't try this at home; it is not a good idea. In the 1960s and 1970s few if any science departments offered students any health and safety training at all.

2 Romaine Malenfant, "Risque et gestion du social: le retrait de l'activité professionnelle durant la grossesse," *Recherches Sociographiques* 39 (1998): 39–57; Romaine Malenfant, Anne Renée Gravel, Normand Laplant, and Robert Plante, "Grossesse et travail: Au-delà des facteurs de risques pour la santé," *Revue multidisciplinaire sur l'emploi, le syndicalisme et le travail* (2011): 50–72. www.remest.ca/documents/MalenfantREMESTVol6no2.pdf.

3 Robert Plante and Romain Malenfant, "Reproductive Health and Work: Different Experiences," *Journal of Occupational and Environmental Medicine* 40 (1998): 964–68.

4 Marc Renaud and Geneviève Turcotte, *Comment les travailleuses enceintes voient leur travail, ses risques et le droit au retrait préventif. Rapport de recherche* (Montréal: Groupe de recherche sur les aspects sociaux de la prévention, Université de Montréal, 1988). Renaud was the man who started the CQRS grants for community research partnerships (see preface).

5 Suzanne Cohen, "Opération du sauvegarde du programme: pour une maternité sans danger," *Prévention au Travail* (1998, Oct-Nov-Dec): 7–9.

6 Agathe Croteau, Sylvie Marcoux, and Chantal Brisson, "Work Activity in Pregnancy, Preventive Measures, and the Risk of Preterm Delivery," *American Journal of Epidemiology* 166 (2007): 951–65; Agathe Croteau, Sylvie Marcoux, and Chantal Brisson, "Work Activity in Pregnancy, Preventive Measures, and the Risk of Delivering a Small-for-Gestational-Age Infant," *American Journal of Public Health* 96, no. 5 (2006): 846–55.

7 This report had only a slight effect because Dr. Croteau's articles were published in two of the most respectable public health journals.

8 Katherine Lippel, *Workers' Compensation for Asbestos Related Disease in Five Canadian Provinces* (Ottawa: Canada Research Chair in Occupational Health and Safety Law, 2010).

9 In my field, *New Solutions: A Journal of Environmental and Occupational Health Policy* does a great job.

10 Like Spirit of 1848 in the U.S., a listserv for public health scientists, www.spiritof1848.org/. In Canada, universities are currently setting up a network for community-based research, www.communitybasedresearch.ca/.

11 Agathe Croteau, Robert Plante and colleagues showed that those who were exposed to risks and took precautionary leave had better pregnancy outcomes than those who did not.

Index

Note: "n" after a page number indicates an endnote; "nn" after a page number indicates two or more consecutive endnotes.